Girls' Secondary Education in the Western World

SECONDARY EDUCATION IN A CHANGING WORLD

Series editors: Barry M. Franklin and Gary McCulloch

Published by Palgrave Macmillan:

The Comprehensive Public High School: Historical Perspectives
By Geoffrey Sherington and Craig Campbell
(2006)

Cyril Norwood and the Ideal of Secondary Education
By Gary McCulloch
(2007)

The Death of the Comprehensive High School?:
Historical, Contemporary, and Comparative Perspectives
Edited by Barry M. Franklin and Gary McCulloch
(2007)

The Emergence of Holocaust Education in American Schools
By Thomas D. Fallace
(2008)

The Standardization of American Schooling:
Linking Secondary and Higher Education, 1870–1910
By Marc A. VanOverbeke
(2008)

Education and Social Integration:
Comprehensive Schooling in Europe
By Susanne Wiborg
(2009)

Reforming New Zealand Secondary Education:
The Picot Report and the Road to Radical Reform
By Roger Openshaw
(2009)

Inciting Change in Secondary English Language Programs:
The Case of Cherry High School
By Marilee Coles-Ritchie
(2009)

Curriculum, Community, and Urban School Reform
Barry M. Franklin
(2010)

Girls' Secondary Education in the Western World:
From the 18th to the 20th Century
Edited by James C. Albisetti, Joyce Goodman, and Rebecca Rogers
(2010)

GIRLS' SECONDARY EDUCATION IN THE WESTERN WORLD

FROM THE 18TH TO THE 20TH CENTURY

EDITED BY
JAMES C. ALBISETTI, JOYCE GOODMAN, AND
REBECCA ROGERS

palgrave
macmillan

The front cover image, "Miss Caroline Coignou showing girls a toad," is reproduced courtesy of Manchester High School for Girls Archive (www.mhsgarchive.org/index.php).

First published in 2010 by ✓
PALGRAVE MACMILLAN®
in the United States—a division of St. Martin's Press LLC,
175 Fifth Avenue, New York, NY 10010.

Where this book is distributed in the UK, Europe and the rest of the world, this is by Palgrave Macmillan, a division of Macmillan Publishers Limited, registered in England, company number 785998, of Houndmills, Basingstoke, Hampshire RG21 6XS.

Palgrave Macmillan is the global academic imprint of the above companies and has companies and representatives throughout the world.

Palgrave® and Macmillan® are registered trademarks in the United States, the United Kingdom, Europe and other countries.

ISBN: 978–0–230–61946–3

Library of Congress Cataloging-in-Publication Data is available from the Library of Congress.

A catalogue record of the book is available from the British Library.

Design by Newgen Imaging Systems (P) Ltd., Chennai, India.

First edition: May 2010

10 9 8 7 6 5 4 3 2 1

Printed in the United States of America.

Transferred to Digital Printing in 2011

Contents

Illustrations

Tables

Figure

Note on Cover

Miss Caroline Coignou showing girls a toad (front cover). Source: Manchester High School for Girls Archive (www.mhsgarchive.org)

Caroline Coignou taught botany, nature study, geology, physiography, chemistry, natural science and hygiene and was responsible for the "control of laboratories, greenhouses and the school garden" at Manchester High School in England between 1895 and 1910. She subsequently became an Inspector of Schools in the West Riding of Yorkshire.

Series Editors' Foreword

Among the educational issues affecting policy-makers, public officials, and citizens in modern, democratic, and industrial societies, none has been more contentious than the role of secondary schooling. In establishing the Secondary Education in a Changing World series with Palgrave Macmillan, our intent is to provide a venue for scholars in different national settings to explore critical and controversial issues surrounding secondary education. We envision our series as a place for the airing and resolution of these controversial issues.

More than a century has elapsed since Émile Durkheim argued the importance of studying secondary education as a unity, rather than in relation to the wide range of subjects and the division of pedagogical labor, of which it was composed. Only thus, he insisted, would it be possible to have the ends and aims of secondary education constantly in view. The failure to do so accounted for a great deal of the difficulty with which secondary education was faced. First, it meant that secondary education was "intellectually disorientated," between "a past which is dying and a future which is still undecided," and, as a result, "lacks the vigor and vitality which it once possessed."[1] Second, the institutions of secondary education were not understood adequately in relation to their past, which was "the soil which nourished them and gave them their present meaning, and apart from which they cannot be examined without a great deal of impoverishment and distortion" (10). And third, it was difficult for secondary school teachers, who were responsible for putting policy reforms into practice, to understand the nature of the problems and issues that prompted them.

In the early years of the twenty-first century, Durkheim's strictures still have resonance. The intellectual disorientation of secondary education is more evident than ever as it is caught up in successive waves of policy changes. The connections between the present and the past have become increasingly hard to trace and untangle. Moreover, the distance between policy-makers on the one hand and the practitioners on the other has rarely seemed as immense as it is today. The key mission of the current series of books is, in the spirit of Durkheim, to address these underlying dilemmas of secondary education and to play a part in resolving them.

Girls' Secondary Education in the Western World: From the 18th to the 20th Century, edited by James Albisetti, Joyce Goodman, and Rebecca Rogers, contributes to this

---

1. Émile Durkheim. *The Evolution of Educational Thought: Lectures on the Formation and Development of Secondary Education in France* (London: Routledge and Kegan; 1938/1977), 8.

mission through its widely ranging analyses of secondary education for girls as it has developed in many countries around Europe and in America over the past two hundred years. It approaches this theme principally through investigations of the separate national histories involved, including those of Great Britain, Ireland, France, Germany, Austria, Italy, Spain, Portugal, the Netherlands, Belgium, Scandinavia, Bulgaria, Russia, and the United States. These provide systematic treatments of the relevant historiographies, the main changes since the eighteenth century, developments in a range of educational settings, the relationship to secondary education for boys, the extent of coeducation, and contributions to girls' secondary education in the overseas colonies.

The analyses of different systems in the Western world demonstrate the distinctive ways in which secondary education for girls has developed historically in relation to different social, cultural, religious, and political contexts. They also reveal international and comparative frameworks for the understanding of issues that have affected girls' secondary education over the past two hundred years. For example, they assess the extent to which secondary education for girls has constituted a conservative or a radical element in social and political change over the longer term. Moreover, they provide a means of tracing the experience of girls' secondary education for the pupils, teachers, and administrators involved, no less than the policies and ideologies that helped to shape it. In such ways, this volume highlights the similarities and common patterns in the historical development of secondary education for girls as well as the wide range of variation and difference. It thus affords a unique insight into the ways in which European patterns of development in this area have compared with those in America, and indeed the nature of the spread of girls' secondary education across national borders.

Girls' Secondary Education in the Western World is the tenth volume to be published in our series. It exemplifies well the combination of social, historical, and comparative approaches to secondary education that we have sought to emphasize throughout, and is the first to focus on secondary education for girls. As we see the trajectory of the series advancing during the next few years, our intent is to seek additional volumes that bring these issues still further to the attention of studies in secondary education.

BARRY FRANKLIN AND GARY MCCULLOCH

Acknowledgments

We have incurred a number of debts to individuals and to institutions in writing this book. We are extremely grateful to our contributors for their hard work in producing their chapters, and thank especially those whose first language is not English. As editors, we much appreciated your patience as we grappled with the particularities of education systems across Europe and their provision for the secondary education of girls.

Several contributors to this volume had the opportunity to present early versions of their chapters in a symposium for Network 17 (Histories of Education) at the European Educational Research Association Conference (2008) in Gothenburg. Many thanks to all those who took part in the symposium and to all who provided helpful comments at this early stage of the work.

Jim is very grateful to Rebecca and Joyce for involving him fully in what began as their joint project. He wishes to thank the College of Arts and Sciences and the Department of History at the University of Kentucky for funding most of his trip to the EERA meeting in Sweden.

Joyce wishes to thank the University of Winchester Research Grant Scheme for support for the three-year research project, "The International Mind and the Education of Women and Girls," on which she is engaged with Andrea Jacobs from the Centre for the History of Women's Education at Winchester, which funded attendance at the Gothenburg conference. This project acted as a catalyst for her current research on imperial, European, and transnational aspects of girls' secondary education, which is reflected in this book.

Rebecca was especially privileged to have a research leave from the Université Paris Descartes, which allowed her to spend time on this project. Heartfelt thanks to the colleagues who supported this leave and especially the research lab, the Centre de Recherches sur les Liens Sociaux (CERLIS), which funded the trip to Gothenburg, and supports so willingly research in the history of girls' education.

We would like to thank the series editors Gary McCulloch and Barry Franklin for their enthusiasm to include a book on girls' secondary education in their series, "Secondary Education in a Changing World," Amanda Johnson, Julia Cohen, and Samantha Hasey for their support during the writing phase, and Andrea Jacobs for her efficiency in organizing the final submission to Palgrave.

Previous Publications

By James C Albisetti
Schooling German Girls and Women (1989)
Secondary School Reform in Imperial Germany (1983)

By Joyce Goodman
Social Change in the History of British Education (2008, with Gary McCulloch and William Richardson)
Women and Education,1800–1980 (2004, with Jane Martin)
Gender, Colonialism and Education, the Political Experience of Education (2002, with Jane Martin)
Women, Educational Policy-Making and Administration in England. Authoritative Women since 1800 (2000, with Sylvia Harrop)

By Rebecca Rogers
From the salon to the Schoolroom: Educating Middle-Class Girls in Nineteenth-Century France (2005); *Les bourgeoises au pensionnat. L'éducation féminine au XIXe siècle* (translation, 2007); *La mixité dans l'éducation : enjeux passés et présents* (2004, editor)
Les espaces de l'historien. Etudes d'historiographie (2000, with Jean-Claude Waquet and Odile Goerg)
Les Demoiselles de la Légion d'honneur. Les Maisons d'éducation de la Légion d'honneur aux dix-neuvième siècle (1992, 2nd edition 2006)

Girls' Secondary Education in the Western World: A Historical Introduction

James C. Albisetti, Joyce Goodman, and Rebecca Rogers

Interest in comparative study of girls' secondary education in Europe dates back at least to 1884, when Theodore Stanton published *The Women Question in Europe*, a series of essays by feminists from various countries, all of whom highlighted educational issues in their judgment of women's status in their homelands. Two decades later, Helene Lange and Gertrud Bäumer also drew on numerous women, and some men, for chapters on individual countries in the volume of their five-volume *Handbook of the Women's Movement* devoted to education. Shortly thereafter, Käthe Schirmacher similarly placed great stress on educational advances in her sweeping worldwide survey of *The Modern Woman's Rights Movement*.[1] Interest in the link between girls' education and women's emancipation continued into the interwar period. In 1934, a moderate feminist Hungarian teacher named Amélie Arato published a much more detailed but less historical comparative study of the current state of girls' secondary education in Europe. In addition to providing the curricula and timetables for schools in many countries, Arato also discussed the widely differing practices with regard to coeducation and the role of men and women teachers in girls' schools, themes echoed in many chapters of this volume.[2]

Not surprisingly, the early days of modern scholarship in women's history placed education on the historical agenda, notably with the publication in 1978 of Phyllis Stock's *Better Than Rubies: A History of Women's Education*.[3] The explosion of research in the succeeding years, however, soon moved far beyond Stock's synthesis. In the three decades since then, no similar survey has appeared, although, as the bibliography to this volume suggests, scholars have produced comparative studies of limited topics, or of two or three countries, over a limited period of time.

Our collection, structured around national chapters, brings to the reading public a wealth of information concerning the individuals who argued for girls' education, the place and role of women as teachers and administrators within school systems, and the nature of the female schooling experience. Alongside the relatively familiar stories of British, French, or German girls' secondary education, readers will discover the weight of conservative Catholic messages in the histories of Portugal, Spain, Italy, Belgium, Austria, and Ireland; but they will also learn the ways Catholic and Protestant missionaries helped establish serious models of girls' education in Greece and the Balkan peninsula. Very different patterns of development emerged within the northern European countries—the Netherlands and Scandinavia—thanks to the influence of Protestantism and a greater receptivity to coeducation; while the history of girls' access to secondary education in Bulgaria and Russia highlights the ways nation-building and politics penetrated schoolrooms affecting the experiences of both girls and boys. Certain countries remain outside the purview of this volume and still await a scholarly treatment in English. These include Switzerland, famous for the early admission of (mostly foreign) women to universities, but with schools that varied greatly among its twenty-two cantons.[4] Readers interested in the more recent nation-states of central Europe (Poland, Hungary, Romania or Czechoslovakia) will find references to these countries in the final two chapters that discuss the transnational aspects of girls' secondary education, and in the chapter by Daskalova, but a synthetic presentation of their histories still remains to be done.

The authors of individual chapters were all asked to address a certain number of issues to ensure a thematic coherence to the whole. As a result, each chapter begins with an outline of the historiography followed by a chronological discussion of the major changes in educational offerings beginning in the eighteenth century. Alongside analysis of the national educational system and its evolution, authors highlight significant schools, and important women educators and pedagogues, in order to offer cultural as well as more socioeconomic explanations for the evolutions in girls' secondary education. The chapters pay attention to a variety of educational settings: public institutions funded by cities, regions, or the states; private schools run by individual proprietors, nuns, or educational associations; and even homeschooling when the information is available. Each chapter also addresses the increasing convergence of girls' and boys' secondary schooling and the issue of coeducation. Finally, authors were asked to consider to what extent their country contributed to the emergence of girls' secondary education within the colonies.

This book takes seriously the dialectic between education as a conservative force and as a force for change as expressed in both democratic and authoritarian political agendas across Europe. By examining struggles to learn and to acquire knowledge, the book explores the experience of pupils, teachers, and administrators over the past three centuries and the impact of their struggles on the cultural and economic life of women and nations.

The treatment of these themes depends not only on national histories but also on the existing historiography, the strength of women's history, and the availability of sources. Statistical sources vary tremendously within individual countries. Centralized states, such as France, produced educational statistics from the early nineteenth century, but often left out the private sector and hence the vast majority of girls' schools. In Belgium, triennial reports concerning public education presented before Parliament were published from 1842 onward, containing information about institutions, personnel, student numbers, as well as inspection reports. The British colonial state kept statistics for many British colonies. Educational and missionary associations, individual institutions, and regional or municipal governments all provide grist for the social, political, and institutional aspects of girls' education. Increasingly scholars are also drawing on a wealth of other sources including letters, diaries, petitions, and memoirs that demonstrate how individuals circumvented gendered cultural norms.

Secondary Education: Definitions and Debates

The collection's focus on girls' secondary education represents in part a pragmatic decision to narrow the investigation to specific social and age groups, notably the middle and upper classes and girls between the ages of twelve and eighteen. The wealth of sources concerning secondary education, compared with primary education, in addition to the dearth of comparative analysis, justifies the focus on this level. Individual authors provide definitions of what constituted secondary education in their countries and highlight how the meaning changed over time. Readers will discover the complexities of terminology, as similar names often masked institutional differences. The democratic American high school fascinated European observers but had no equivalent in Europe before the second half of the twentieth century. The discussion about how Europeans defined secondary education in chapter 12 provides an illuminating introduction to national understandings of educational levels. School systems were often finely differentiated by class so that in individual countries contemporaries sent their children to institutions that corresponded to a specific social and educational ethos. In general, the most prestigious and elitist of these institutions—the German *Gymnasium* and the French *lycée*—were not available to girls until the end of the nineteenth century, as was largely the case with the British public schools. Because secondary education carried such a strong class connotation in most countries, it did not always strictly apply to pupils over the age of twelve. In this volume, however, authors focus for the most part on the schooling of adolescents and trace a similar pattern of democratization within educational systems.

If secondary education was associated with the bourgeoisie and elites throughout most of the period under consideration in this volume, the focus of individual chapters is on the gendered characteristics of this secondary system. Many of the contributions show that throughout Europe secondary education was oriented toward the training of "public" men. This vision left little room for middle-class women whose primary mission was assumed to be that of good wife and mother. Why, many contemporaries argued, should such a woman study the classics, or learn mathematics and physics? Instead, girls' secondary education should focus on appropriately feminine subjects, most notably foreign languages, sewing, and painting, in addition to the indispensable religious and moral messages. The emergence of a feminine program that was seen as secondary and its relationship to the masculine system form the backbone of each of the following national chapters?

Commonalities and Differences: A Brief Comparative Discussion

Certain common themes emerge in all the chapters and constitute a relatively familiar narrative for women's historians. For girls of the middle and upper classes, "secondary" education involved lessons in domesticity, although the skills necessary for running a home might vary from country to country. The importance of producing educated mothers extended far beyond national borders, as imperial states carried European representations of motherhood to the colonies. Domesticity implied a range of gendered attitudes that were reflected not just in course content but also in prevailing moral ideals that built upon religious values in all of the European countries. While Italian Catholic boarding schools bore little resemblance to the Danish Lutheran daughters' schools, both sorts of institutions placed religious instruction and values at the core of the educational system. Challenges to this domestic and religious orientation came from both male and female feminists and acquired an increasing audience in the second half of the nineteenth century.

Feminist movements varied within individual countries, but all the chapters testify to how feminist ideas about women's ability to reason circulated in the educational world and encouraged the opening of schools or the organization of campaigns, beginning in the 1860s and 1870s. The first international congresses of women placed women's education at the heart of feminist demands. As early as 1871 the *Association Internationale des Femmes* (International Association of Women) in Geneva proclaimed its goals: "To work for the moral and intellectual advancement of women, for the gradual amelioration of her position in society by calling for her human, civil, economic, social and political rights."[5] As a result, while feminist demands were not always the leading force in providing access to secondary education (in Britain, for example, some headmistresses were supporters of the anti-suffrage movement), they almost always had an impact on the tenor of the debate. No single feminist doctrine about what constituted girls' secondary education emerged, but most countries witnessed debate about its content (should it be identical or not with boys), its location (coeducational or single-sex schools), and its goals (to train mothers, citizens, or professionals).

If the final quarter of the nineteenth century represents a moment of intense educational ferment for middle-class girls, the second moment of widespread change occurred after World War II, as demands for secondary education spread. The chapters highlight repeatedly in different contexts how the democratization of the educational system and the widespread extension of coeducation opened new opportunities for women without erasing the historical weight of gendered visions of femininity and women's relationship to knowledge and the public sphere.

Beyond these broad similarities, however, the different chapters reveal the extent to which national contexts determined the nature of the debate about girls' secondary education and the forms it took. Religion emerges here as one of the common threads whose influence had very different impact on girls' schooling throughout Europe. In Italy, Spain, and Portugal, Catholicism played a very conservative role and nuns did little to improve significantly the quality of girls' educational opportunities. In France, however, liberal currents within Catholicism, as well as competition with an active lay sector, encouraged innovation, notably with respect to teacher training. In countries such as Ireland, Belgium, Greece, and Bulgaria, competing religious associations opened schools and sought to attract a clientele; in the process, they established standards with respect to girls' schooling and helped spread the perception that girls should indeed be sent to schools like their brothers. In Ireland the religious rivalry between Catholics and Protestants provided the impetus for reform, enhanced the academic nature of girls' schooling, and raised girls' examination achievements. In Southeastern Europe, innovation came from Catholic and Protestant missionaries seeking to spread their vision of what constituted serious girls' education. The complexity of these national studies should caution us to associate progressivism with Protestantism and conservatism with Catholicism.

If domestic motherhood constitutes a unifying thread to discussions about girls' education throughout most of the three centuries, in some countries nationalism and the concern to form the patriotic mother played a particularly significant role in the emergence of opportunities for girls. The chapter on Southeastern Europe makes this clear notably with the creation of the first modern women's high school (*Visa zenska skola*) in Belgrade in 1863, where nationalist rhetoric underwrote the curriculum. In territories under the rule of the Habsburg and Russian empires, the politics of cultural nationalism gave pedagogues and feminists a discourse about girls' education that resolutely connected the family and its values with the public sphere. In Finland, for example, the educator Lucina Hagman (1853–1946) participated actively in the national movement from the 1880s in addition to opening the first coeducational school for Finnish-speaking students, Helsinki's Suomalainen Yhteiskoulu, in 1886. In a speech to the Stockholm Feminist Congress of 1897, Hagman predicted that coeducation would ultimately "raise the moral standards in our human societies, build character and [...] without straying from the path Nature has indicated, allow men and women cordially to take one another by the hand and work together at the great common project of perfecting the human species."[6] In certain countries in Western Europe, such as France, Belgium, and Portugal, the politics of anticlericalism more than nation-building influenced the development of secondary schools' for girls. Republicans or liberals, supported by feminist associations, created public institutions for girls in an effort to wrest their education from the hands of the Catholic Church.

If politics and nationalism explain differences between the different countries in their approach to girls' education, the characteristics of national educational systems also played an important role. This is apparent in attitudes toward coeducation among pupils, as the chapter on the Netherlands demonstrates, but also with respect to the sex of school teachers. In France and England, girls' secondary schools developed under the direction of strong women teachers who argued that only women could run schools for girls. This allowed women directors to achieve positions of importance in the educational hierarchy, in contrast to Germany or the Scandinavian countries where men contested women for school headships. As in Scotland, men headed girls' schools in Austria and Russia, and for many years women taught only in the lower grades. Similarly, in Calvinist Geneva the *École secondaire et supérieure des filles* (the secondary and higher school for girls) founded in 1847 established a seven-year program of study, but only men taught within the school.[7] Paradoxically, male teachers and directors encouraged the early development of girls' schools but prevented women from achieving positions of responsibility. In general, throughout Europe, women teachers found it difficult to lead fulfilling lives as both professionals and mothers, so that even when bans on marriage did not exist, the model for the secondary woman teacher was that of the celibate woman, or the nun in Catholic countries.[8]

Access to secondary education for girls took very different routes throughout Europe. In some countries, such as France, Germany, or Belgium, girls' schools emerged alongside an existing male system and shared many of the latter's characteristics, notably a conviction that secondary studies should be "disinterested" and not professional. In Italy and Spain, however, the emergence of more serious studies for girls took a more vocational path via teacher training schools, ostensibly oriented toward training primary schoolteachers. In these countries, girls aspiring to postprimary education took advantage of existing structures that did not challenge gender norms, since throughout Europe education was seen as an acceptable profession for women. In England, opportunities for lower-middle-class girls also came via teacher training, while girls from higher social classes followed a more "liberal" education track. More audaciously in some countries, families used the absence of girls' secondary schools as an argument to place their daughters in boys' secondary schools. This was the case in the Netherlands, as early as 1871, but also in the educationally "backward" Mediterranean countries of Italy, Spain, and Portugal. In Italy, historian Marino Raichich has described this relatively early access to boys' secondary institutions (1883) as one of the "advantages of backwardness."[9]

State of the Field: Orientations

The historiography concerning girls' secondary education varies tremendously from country to country and this is naturally reflected in the following chapters. For some countries, notably England, France, and Germany, broad studies of this level of schooling already exist, although none cover three centuries (see bibliography). Other countries have only begun to explore the complexities of this history, often through the lens of women's history and a focus on specific figures. The recent proliferation of biographical dictionaries undoubtedly has fostered an interest in women pedagogues and educators, although no European equivalent to Linda Eisenmann's *Historical Dictionary of Women's Education* exists to date.[10] The focus on biography, however, has not yet modified durably the dominant narrative in the history of education, which continues to judge girls' schooling in comparison to that of boys without querying how the two developed in tandem and the ways the analytical category of gender might rewrite understandings of boys' schools and masculinity. This no doubt remains a goal for the future.[11]

The volume reflects most strongly the strength of scholarship on the nineteenth and early twentieth centuries. This is when the most important girls' institutions emerged, but recent work on the period of the Enlightenment has shown the importance of debates about girls' education that modern historians need to acknowledge.[12] Renewed interest in the Enlightenment and the early nineteenth century is evident across Europe and merits the type of transnational approach adopted in the last two chapters of this volume. Ideas as well as educational models circulated well before the first international education congresses. The Smolny institute in Russia (founded in the 1760s), for example, drew on the example of its predecessor, the school for noble girls in Saint-Cyr, France (founded in 1686). In Poland, prior to 1795, Enlightenment ideas influenced the Commission for National Education in 1773 that insisted on the need to provide education to both sexes (although it did not create specific institutions for girls).[13]

At the chronological end of the period under investigation, the individual chapters skip quickly over the post-Second World War years when democratization of "middle schools" and coeducation ended the most apparent differences between boys' and girls' secondary education. As the analyses of Germany and Portugal suggests, however, these years of massive change in the educational system merit more careful attention. Up until now sociologists have largely dominated the scholarly production for this period and it is clearly time to integrate a more historical perspective to understand why the end of girls' secondary education, as it was understood, has not eliminated gender inequalities in the educational system.

Despite the postcolonial turn in the English-speaking world, these chapters reveal that much remains to be done on the subject of colonial schools and the ways governments, associations, and women teachers contributed to the civilizing mission. Closer attention to the cultural work of colonization should provide a useful corrective to a history that remains dominated by a British narrative.[14]

The effort to provide a concise synthesis over 300 years for each chapter has led to regrettable absences. The chapters say very little about physical education or girls' classroom experiences, and the material aspects and the physical setting of schooling are not the focus of any sustained analysis. Information about informal education, in general, be it at home or in school, is only briefly mentioned in several chapters, despite the fact that such education was an important aspect of girls' self-development, as biographical approaches have shown. Source material is not always available for such studies, but our hope is that this volume will stimulate others to pursue threads only briefly touched upon. The broad scope of this volume, both in chronological and geographic terms, ultimately should stimulate reflection about the complexity of historical periods and the importance of cross-national discussions and borrowing; it also points in interesting ways to the historicity of national ideas about the curriculum and what was deemed necessary for cultured young European woman to know. By pushing readers to compare across borders and to consider the effects of including girls in the history of individual systems, the book aims to stimulate research conversations and a research agenda for scholars interested in history, education, and women's lives.

Notes

1. Theodore Stanton, ed., *The Women Question in Europe: A Series of Original Essays* (New York: Putnam, 1884 [reprint New York: Source Book Press, 1970]); Helene Lange and Gertrud Bäumer, *Handbuch der Frauenbewegung.*, vol. 3: *Der Stand der Frauenbildung in den Kulturländern* (Berlin: W. Moeser, 1902); Käthe Schirmacher, *The Modern Woman's Rights Movement: A Historical Survey*, trans. Carl Conrad Eckhardt (New York: Macmillan, 1912).
2. Amélie Arato, *L'enseignement secondaire des jeunes filles en Europe* (Brussels: J. Lebèque, 1934).
3. Phyllis Stock, *Better Than Rubies: A History of Women's Education* (New York: Putnam, 1978).
4. For studies from widely different eras, see Ferdinand Zehender, *Geschichtliche Darstellung des öffentlichen Unterrichts für Mädchen in der Stadt Zürich von 1774 bis 1883* (Zurich: Schulthess,

1883); Gottlieb Rothen, *Hundert Jahre Mädchenschule in der Stadt Bern* (Bern: Mädchensekundarschule, 1936); and Chantel Renevey-Fry, ed., *En attendant le prince charmant: L'éducation des jeunes filles à Genève, 1740–1970* (Geneva: SRED, 1997).

5. See Karen Offen, *European Feminisms, 1700–1950: A Political History* (Stanford: Stanford University Press, 2000), especially chapters 5 and 6, here 152–153.

6. Lucina Hagman, "La Coéducation des sexes en Finlande," *Revue de morale sociale* (1899): 42.

7. Renevey-Fry, *En attendant le prince charmant*, 125–141.

8. Leslie Page Moch, "Government Policy and Women's Experience: the Case of Teachers in France," *Feminist Studies* 14 (1988): 301–324.

9. Marino Raichich, "Liceo, università, professioni: un percorso difficile," in *L'educazione delle donne. Scuole e modelli di vita femminile nell'Italia dell'Ottocento*, ed. Simonetta Soldani (Milan: Francoangeli, 1989), 113.

10. Linda Eisenmann, ed., *Historical Dictionary of Women's Education in the United States* (Westport: Greenwood Press, 1998).

11. Ruth Watts, "Gendering the story: change in the history of education," *History of Education* 34, 3 (May 2005): 225–241; and Rebecca Rogers, "The Politics of Writing the History of French Girls' Education," *History of Education Researcher* 80 (November 2007): 136–144.

12. Carol Strauss Sotiropoulis, *Early Feminists and the Education Debates: England, France, Germany, 1760–1810* (Madison, NJ: Fairleigh Dickinson University Press, 2007).

13. Maciej Serwanski, "Les formes de l'éducation des filles nobles en Pologne aux XVIe, XVIIe et XVIIIe siècles," in *L'éducation des jeunes filles nobles en Europe, XVIIe–XVIIIe siècles*, ed. Chantel Grell and Arnaud Ramière de Fortanier (Paris: Presses de l'Université Paris-Sorbonne, 2004), 75–85.

14. For an initial effort in this direction, see Julia Clancy-Smith and Frances Gouda, *Domesticating the Empire: Race, Gender, and Family Life in French and Dutch Colonialism* (Charlottesville and London: University Press of Virginia, 1998). This collection provides a great deal of information about girls' education, but none of the chapters deals explicitly with the subject.

Chapter 1

Class and Religion: Great Britain and Ireland

Joyce Goodman

Girls' secondary education shared similarities but also differed in England, Wales, Scotland, Ireland, and the British colonies. It was inflected by Anglicanism in England, Presbyterianism in Scotland, Nonconformity in Wales, and Catholicism in Ireland.[1] Secondary education (termed "intermediate" in Wales and Ireland) was largely accessed by upper- and middle-class girls, while elementary education was reserved for the working class. Within secondary education, independent fee-paying schools were the preserve of the elite, and endowed and state-maintained grammar schools were attended by middle-class and bright working-class girls (the latter via a scholarship "ladder").[2] In contrast, the Scottish tradition of the "democratic intellect" (providing a path to University for poor boys), did not lead state-maintained schooling to be sharply differentiated by social class into elementary schooling for the working class and secondary for the middle class. Scottish education was largely mixed, while Wales introduced dual schools (a girls' side and a boys' side under one roof) in 1889. In England single-sex girls' secondary education predominated until comprehensive schools became widespread in the 1970s, although in Ireland single-sex education continues to remain more prevalent than in England.

The majority of research on girls' education has focused on education and constructions of middle-class femininities in England.[3] Recent scholarship has attended to Enlightenment debates, to education for girls prior to the nineteenth-century women's movement,[4] and to Welsh, Scottish, and Irish girls' schooling.[5] Research covers the contradictory nature of girls' education, as schools promoted both domestic notions of femininity and masculine educational standards; gendered cultures of schooling, teaching, and knowledge; social Darwinistic and eugenic ideas of national efficiency;[6] and gendered technologies of Self.[7] Cultural, visual, material, auto/biographical, oral, and transnational methodologies are enriching understandings of girls' experience of schooling; of women who taught, administered, engaged in policymaking, or agitated for change in the education of girls; and of girls' education and empire.[8]

The chapter divides discussion on the topic into four periods and ends with a brief discussion of colonial education. The first period examines public/private schooling for girls prior to 1868; the second, from 1868 to 1914, discusses the rise of girls' high schools, convent education, and educational opportunities for working-class girls; the third illustrates that "difference" increasingly defined girls' education in the interwar years; the fourth looks at girls' schooling since 1945.

Public and Private Schooling Prior to 1868

Women's education formed an important part of seventeenth- and eighteenth-century debate.[9] Among those who viewed the separation of mind and body as grounds to assert intellectual equality was the philosopher and educationist Mary Astell (1666–1731), who advocated an all-women's community, where women could spend their time in study and contemplation.[10] Others such as Priscilla Wakefield (1750–1832) drew on Locke's *Essay Concerning Human Understanding* (1691) to argue that the origins of ideas in the sensations received from the external world could be combined through the association of ideas by women as well as men. The "bluestocking" circle saw women's learning as part of a domesticated culture capable of refining aristocratic excess, while Scottish Enlightenment "stage" theorists placed women's education central to the "progress" of society.[11]

Whether education should be private (home-based) or public (school or convent-based) was debated for boys as well as girls. Locke saw the public grammar school as detrimental to boys' morals and argued that boys' education should be conducted in the private (domestic) sphere; but the view of headmaster Vicesimus Knox (1750–1832) predominated: the domestic environment endangered boys' manliness, whereas the school offered the discipline for a grounding in the classics to form boys' virtue. Public forms of schooling that benefited boys were thought inappropriate for girls, as was mixing social classes in girls' boarding schools and, during wars with France, French convent education.[12] Rousseau's portrayal of Sophie in *Emile* (1762), his view of the male and female mind as different, and his argument that schooling should relate to a girls' future role as wife and mother supported the belief that girls should be educated at home. Many eighteenth- and nineteenth-century mothers and governesses provided a home education for girls, who might also attend one or more schools of variable standards, where the emphasis was on English grammar, conversational French, and the "accomplishments" of music, art, drawing, and dancing.

Proprietors' descriptions of girls' schools in domestic terms concealed complex and contradictory meanings surrounding girls' education. Michèle Cohen illustrates how the "superficiality" associated with girls' schooling built on understandings of "method" associated with order and system that originated in humanist ideas from the 1510s. These had been applied to Latin, which was learned through a graded system employing repetition, which was thought to require discipline. Narrowness in the boys' curriculum and the ability to be absorbed by a single subject was valued for "thoroughness" and training the mind. It was contrasted with the "superficiality," elegance, and politeness seen as hallmarks of girls' education. English grammar was taught in girls' schools because of its association with the Latin grammar taught to boys. French was regarded as an accomplishment and associated with frivolity because it was taught through conversation, rather than, like Latin, through grammar and translation. The "diffusion" of the girls' broader curriculum, which included history and botany before they were taught to boys, denoted a lack of authority and was mapped onto a "natural" mental difference that upheld women's supposed lack of fitness for higher study.[13]

Both Hannah More (1745–1833) and Mary Wollstonecraft (1759–1797) thought women's education lacked method and system and cramped the mind.[14] For More, education was to make good daughters, wives, members of society, and Christians. More drew on Evangelical theology and Lockean psychology to argue that girls' trivial education encouraged vanity and stifled moral growth. While not advocating autonomy for women, she advised "dry tough reading" to inculcate habits of study, alongside traditionally "masculine" subjects, a combination she thought conducive to virtue.[15] Wollstonecraft argued for an education to foster intellectual and human autonomy while not displacing responsibilities of motherhood. Using the language of radical republicanism and the ideas of Locke, she critiqued girls' education, and Sophie's education in Rousseau's *Emile*, for a failure to develop reasoning, and for preparing women to become mistresses of men. Wollstonecraft linked a rational education for women to the well-being of the State, the possibility of woman's financial independence, and the duty of woman to herself.[16]

By the mid-nineteenth century, at least half of all middle-class girls in England attended a private school. These ranged from fashionable schools emphasizing the accomplishments, through schools in France where girls "finished" a polite education, to the academically sound schools of the Unitarians and Quakers. The curriculum at Susannah Corder's Quaker girls' school in 1824 included English grammar, arithmetic, geography, astronomy and the use of the globes, ancient and modern history, elements of mathematics, of physics or experimental philosophy, chemistry, natural history, French, and needlework. Latin, Greek, German, Italian, and drawing were provided at extra charge.[17]

Commercial directories and advertisements illustrate schoolmistresses operating schools as businesses and emphasizing a familial atmosphere to reconcile their commercial position with ideals of femininity.[18] Some schoolmistresses and governesses experienced "status incongruence" from working in the homes of others. Other families prepared girls for governessing and teaching through attendance at schools like the Godolphin School, Salisbury, or elementary training colleges like the Borough Road.[19] From 1846, elementary teacher training was reorganized via the pupil-teacher system, which drew on schemes in Holland and Switzerland. Preliminary training consisted of supervised teaching in a school and academic instruction from the head teacher, after which the pupil-teacher competed for the Queen's scholarship to attend a training college.[20] This provided lower-middle-class girls with a secondary-type education.[21] From 1855 the sisters of Notre-Dame de Namur organized pupil-teacher training for Catholic girls at Mount Pleasant, Liverpool. The Society of the Holy Child Jesus (1856), the Society of the Sacred Heart (1873),[22] and many teaching congregations eventually followed.[23]

In the mid-nineteenth-century, calls for more systematic education, formal training, and certification for middle-class governesses led Christian Socialist

Frederick Maurice (1805–1872) to open Queen's College in London (1848) and the Unitarian Elizabeth Jesser Reid (1789–1866) to found Bedford College (1849).[24] Influential headmistresses Frances Buss (1827–1894) and Dorothea Beale (1831–1906) attended Queen's College. The growth of examination opportunities for girls was also significant. The College of Preceptors (1846) developed examinations for middle-class teachers and included girls in examinations in 1851. Girls' examination successes enabled women to exhibit their competence as teachers to prospective parents. Emily Davies (1830–1921), who would found Girton College at Hitchin in 1869, headed the successful campaign to open the Cambridge Local Examinations to girls from 1863.[25] Competition amongst girls and "emulation" remained a source of anxiety around feminine modesty, however.[26]

Demographic concerns about "surplus" women; agitation around girls' education from groups at the center of the nineteenth-century women's movement, like the Langham Place Circle, including at meetings of the Social Science Association (established 1857); and petitioning by women bore fruit when the Schools Inquiry Commission (1868) included girls' schools in its remit.[27] The Commission was the first government commission to take evidence from women in person. Emily Davies, Frances Buss, Dorothea Beale, and teacher and feminist campaigner Elizabeth Wolstenholme (1833–1918) were expert witnesses. Cohen demonstrates that Commissioners framed their investigation of girls' education through eighteenth-century discourses of "superficiality," "thoroughness," and "method." Commissioners found no "natural" inferiority in the mental powers of girls and no differences in their intellectual capacity and aptitudes, but translated fears of emulation for girls, and what they saw as girls' over-eagerness to learn, into notions of health that were deemed to lead to "overstrain." This was used to explain away the finding (also apparent from College of Preceptors' examinations) that girls could outperform boys, which was thought to pose a danger for girls' health.[28]

From the eighteenth century onward, dynamics and discourse around meanings of public and private schooling differed in Scotland and England.[29] Specialist teachers in Scotland taught the daughters of the wealthy at home, but girls educated at home and those in boarding schools also attended private classes offered by teachers at their own house or school, where they were taught in groups of up to twenty. The term "private" described classes devoted to a single subject, rather than one-to-one tuition.

Scottish Enlightenment phrenologists critiqued private "classes" and "accomplishments" for inhibiting the development of a curriculum based on pupils' physiological and psychological stages of development. In the 1820s and 1830s, innovative day schools for pupils aged 11–15 were established in Edinburgh and Glasgow with carefully organized curricula extending over four years. Lindy Moore notes the contrast in size between these schools (80 girls attended the Scottish Institute for the Education of Young Ladies in 1836 and 140 by 1845–1846) and the small domestic middle-class girls' schools in England catering for about 24 pupils. The weekly curriculum of the Edinburgh Collegiate Institution comprised seven hours English, object lessons, elocution, belles lettres, geography, history, and mathematics (usually arithmetic), four hours each of French and German, four for writing and arithmetic, and two each for Italian, singing, and dancing. Latin and Greek were offered as alternatives to modern languages. From the 1830s and 1840s, elementary science was incorporated into the curriculum and was also provided in specific lectures or private "classes," where girls might conduct experiments in practical chemistry.

The pupil-teacher system provided a route for Scottish women into teaching but was resented as an alien English imposition undermining the Scottish tradition of university education for teachers. Moore notes that in England teachers struggled to achieve recognition of their profession in the nineteenth century, but in Scotland they fought to prevent what was already a profession from dilution through the admission of teachers with no University experience, including women prior to 1892. She argues that one consequence of men managing the prestigious girls' schools was that pronouncements in the 1830s and 1840s about girls' intellectual capacities were made to the powerful upper-middle-class of Edinburgh and Glasgow by influential men, often involved in the professional and pedagogical issues of the day. The Scottish tradition of male graduate secondary teachers at a time when women were not able to graduate effectively blocked women from establishing prestigious girls' schools in Scotland and resulted in no powerful headmistresses becoming a visible part of the Scottish women's movement.

Moore concludes that a long-held view that the education of girls in Scotland was more "advanced" than that of England led to a general complacency about the standard of schooling for middle-class girls and delayed the development of any strong public opinion regarding the need for improvements until the 1870s, while the assumption that schools were best managed and taught by men lasted into the twentieth century.

Girls' High Schools, Convent Education, and Working-Class Girls' Secondary Education, 1868–1914

The Schools Inquiry Commission found only fourteen endowed grammar schools for girls in England and Wales compared with 820 endowed schools for boys and called for the establishment of new girls' schools based on the model of Frances Buss' North London Collegiate School, where students were classified by stage of instruction.[30]

When progress in establishing new girls' endowed schools appeared slow, Maria Grey (1816–1906) (née Shirreff) and her sister Emily Shirreff (1814–1897), Lady Henrietta Stanley (1807–1895) and Mary Gurney (1836–1917) established The Girls' Public Day School Company (GPDSC), later Trust, in 1872 as an offshoot of the Women's Education Union for the Education of Girls of All Classes Above the Elementary (1871). The GPDSC founded thirty-eight schools following Buss' model by 1901. It also opened the Maria Grey Training College for women secondary school teachers. The Church Schools Company (1883) founded twenty-four schools by 1896. Ninety new endowed grammar schools for girls were established between 1889 and 1900. Larger reformed boarding schools, often modelled on Cheltenham Ladies College, were created as the counterparts of the leading boys' schools of the time.[31]

Two broad groups worked to reform girls' education in England. The "uncompromising" insisted that girls should study the same subjects as men and sit for the same examinations, while the "separatists" worked to improve girls' education but did not wish to replicate the education offered to boys. Women reformers were associated with the women's movement but also included those, like Lucy Soulsby, first headmistress of Oxford High School (GPDSC), who espoused antifeminism and worked for reform for woman within her sphere.[32] Sara Delamont and Carol Dyhouse argue that reformers redefined but did not reject the Victorian concept of femininity. Schools founded in the last quarter of the nineteenth century aimed to provide a liberal education that included classics, mathematics, and science. Attendance in the morning enabled girls to learn domestic skills at home in the afternoon, and only a small amount of needlework was taught at school. Sport, medical inspection, and dress reform were introduced to counteract fears that education was undermining girls' health and to redress critiques from medics that there was "sex in mind and body."[33] Up to the First World War, girls whose families wanted them to marry well, however, still faced a home education, and periods at day schools and boarding schools of variable quality.

In contrast to England, where reformers campaigned for a share in the educational advantages of boys, the campaign for girls' secondary education in Wales formed part of a wider struggle for secondary and higher education for both sexes.[34] Gareth Elwyn Jones shows that models developed in England influenced reform of girls' schooling in Wales. Welsh campaigners had strong links with English headmistresses such as Buss and Beale and the English women's movement. The Association for Promoting the Education of Girls in Wales (1886–1901), the Cymmrodorion Society, and the National Eisteddfod provided platforms for women with English links to debate girls' education. Elizabeth Hughes (1851–1925) (who became head of the Cambridge Training College for women secondary teachers in England), Dilys Glynne Jones (1857–1932) (who gathered evidence for the Bryce Commission), and Sophie Bryant (1850–1922) (who followed Buss as head of the North London Collegiate School), raised questions about the different conditions pertaining to girls' education in England and Wales.

The Departmental Committee to Inquire into Intermediate and Higher Education in Wales (Aberdare Commission, 1881) highlighted the inadequate provision for girls in Wales, where there were only two endowed grammar schools for girls educating 263 pupils. Bryant, Hughes, and Frances Hoggan (1843–1927), who had studied medicine at Zürich University from 1867, argued for Welsh language teaching, for examinations controlled by the Welsh, and for the conditions of Wales to be considered, rather than direct borrowing of an English system. An intermediate tier of education was introduced for pupils up to the age of eighteen. This provided a route for working-class and lower-middle-class girls to the examinations of the Central Welsh Board. Girls' education followed that of boys, with a curriculum of Latin, Greek, Welsh and English language and literature, modern languages, mathematics, nature study, and applied science.

The introduction of coeducation and dual schools was a major difference between the implementation of the Welsh Intermediate Education Act (1889) and the situation in England. In England, coeducation had been considered by the Schools Inquiry Commission and the Bryce Commission (1895), and had been adopted in the elementary sector and by a small number of "progressive" schools only. By 1900, the ninety-three Welsh intermediate schools included forty-three dual schools, twenty-two boys' schools, twenty-one girls' schools and seven mixed schools. By 1914 there were twenty-three girls' schools, twenty-two boys' schools, forty-six dual schools, and nine mixed schools. Pupil numbers, cost, and attitudes to coeducation were key drivers in the system of Welsh intermediate education, but the subordinate position of the head teacher of the girls' department in the dual school was to prove an issue of contention for women.

Links to the English campaign for girls' education and to the women's movement in England were also apparent in the development of intermediate education for girls in Ireland. Campaigners for Protestant Irish girls' education included Isabella Tod (1836–1896), who founded the Ladies Institute Belfast (1867), Margaret Byers (1832–1912) of the Ladies Collegiate School, Belfast (1859), and Anne Jellicoe (1823–1880) of Alexandra College, Dublin.[35] Alexandra College followed the pattern established at Queen's College, London. The curriculum included theology, arithmetic, algebra, natural science, mental and moral sciences, and Latin, in addition to modern languages, history, drawing, and music. Alexandra College played a part in the inclusion of girls in the Intermediate Education Act (1878) and in the opening of local examinations for women by Trinity College (1869). It became a model for girls' academic education in Ireland.

Deirdre Raftery notes that by 1899 girls were performing well in local examinations in Latin, Greek, and English, although English, French, and domestic economy were the subjects taken by the greatest number. Until 1901, examination results from Protestant girls' schools outstripped those of the Catholic convent boarding schools, where politeness and deportment were highly valued. In the early days of the convent schools, the language of instruction was often French, reflecting the foundation of many orders outside Ireland, including Loreto (Institute of the Blessed Virgin Mary), the Dominicans, the Ursulines, the Sisters of St Louis, and the Society of the Sacred Heart, in addition to Irish Orders such as the Mercy and Holy Faith Sisters. Between 1838 and 1878 a convent education generally consisted of languages, use of the globes, writing, arithmetic, and the accomplishments: drawing, music, and needlework. After pressure from parents led to the inclusion by the 1890s of Latin and mathematics, the performance of Catholic girls in the intermediate examinations was noteworthy.

Convent schools in England also largely stressed religion until pressure from parents increased preparation for the Oxford and Cambridge local examinations.[36] Catholic day schools and boarding schools catering to demand from the growing wealthy Catholic middle-class were established by congregations originating in France and Belgium. By 1887, thirty-two of the sixty-two congregations teaching girls in England had come from France, and five from Belgium. The Sisters of Notre-Dame de Namur (SND) gained a reputation for the quality of their education, which included advanced science, shorthand, typing, art, music, and French by the end of the nineteenth century.

At the start of the twentieth century, the majority of schools providing secondary education for middle-class girls in England were single-sex. From the 1890s, bright working-class girls could gain a secondary-type education in higher grade schools provided by the school boards on the "top" of the elementary system and in the increasingly academic pupil-teacher centers.[37] Both were organized on a mixed model that segregated education for girls and boys in the same building. Higher grade schools were large, with well-qualified staff and good equipment, and girls received a distinctive scientific and technical curriculum. In 1895, all pupils at the Leeds Central Higher Grade School studied English, history, geography, Latin, French, mathematics, at least two sciences (comprising from 11.5 to 14.5 hours according to age, out of a twenty-seven-hour week), drawing, manual instruction, and physical education, with religious instruction, German, shorthand, bookkeeping, geometry, and dressmaking included at various points during the four-year course.[38]

In Scotland, where the tradition was not to separate state schooling into elementary and secondary divisions by social class, the Scottish Education Department added advanced departments to some mixed elementary schools and also created mixed higher grade schools. In cities like Glasgow, where middle-class parents sought a serious secondary education but were reluctant to send daughters to mixed schools, the School Board moved closer to the English position by developing single-sex high schools for girls. This increased social class divisions between girls through the separation of higher from elementary subjects.[39]

The 1902 Education Act in England and Wales established municipal secondary schools, but abolished the higher grade schools and transformed the training of elementary teachers.[40] To encourage more middle-class students into elementary teaching, intending teachers were relocated to the secondary sector up to the age of sixteen, after which they were supported in the secondary school for a further year by a scholarship, and spent a year in school as a practice teacher before proceeding to training college.[41] The establishment of a grammar school scholarship ladder provided for a small number of bright working-class girls to attend the municipal secondary schools, but in England the majority lost a significant route into secondary education with the demise of the higher grade schools. In Wales, schools founded under the Welsh Intermediate Act had been created as an amalgamation of grammar and higher grade models for a predominantly working-class society with a growing middle-class element. They remained after 1902 but took on the humanistic curriculum of the secondary school. The structure of the Welsh system was modified as new municipal schools, both single-sex and mixed, were founded under the 1902 Education Act.[42] In Scotland, higher grade schools only became part of the common secondary school structure in the 1920s, and access to secondary education in Scotland remained remote for many girls in rural areas between 1900 and 1930.[43]

After the 1902 Education Act the number of secondary schools in England grew rapidly from 685 in 1906 to over 1000 by 1913. Whereas in 1897 only 20,000 girls were being educated in recognized secondary schools, over the next twenty years the numbers grew to 185,000. The curriculum provided a liberal education that was increasingly differentiated by gender. Fears of physical deterioration and national fitness in the wake of the Boer War and eugenic stress on maternity and motherhood led to attacks on the academic curriculum of girls' schools. For eugenicists, middle-class women were key to national efficiency because of their supposed hereditary superiority, but needed encouraging toward maternity and training for motherhood. Practical training for domestic duties became a requirement for secondary schools in receipt of government grants in 1905. By 1908, girls over the age of fifteen were permitted to substitute domestic subjects for science and for mathematics other than arithmetic. Between 1909 and 1913, the Board of Education referred domestic teaching in secondary schools to two committees, which considered experiments relating science and housecraft. Some schools introduced domestic subjects, but many academic girls' schools, including those of the GPDST, saw preparation for domesticity and a liberal education as incompatible, although they continued to stress propriety of dress and impropriety of contact with boys.[44]

Eugenic arguments about the importance of racial fitness and purity were disseminated in England through the work of American G. Stanley Hall, who saw adolescence as a time of instability when girls needed special protection from serious study. Hall drew on notions of race-recapitulation to coordinate childhood and youth with the development of the race, and to argue that evolution produced greater differentiation between the sexes. Like Herbert Spencer, he thought that women might succeed in an intellectual education, but at the price of sterility and ill-health. He prescribed a curriculum that excluded ancient classical languages and advanced mathematics and science. To ensure the future of the race and of civilization, girls were to be taught housekeeping and motherhood in buildings replicating the ideal home and to undertake child study, using laboratory methods.[45]

Interwar Girls' Schooling and "Difference"

Hall's views resonated into the interwar period, when his writings became linked with readings derived from Freud, Jung, and Adler and with social Darwinistic and eugenic concerns in the work of medical authorities, evolutionary thinkers, social psychologists, and some headmistresses.[46] His ideas informed the 1923 Board of Education *Report of the Consultative Committee on the Differentiation of the Curriculum for Boys and Girls Respectively in Secondary Schools.* This aimed to determine whether greater gender differentiation was required in the secondary curriculum. The Committee's evidence and findings were contradictory. Psychologists suggested that the similarities between girls' and boys' attainment demanded a common curriculum, whereas differences in emotion and temperament demanded curricular differentiation. Women as well as men teachers thought that emotional and temperamental differences between boys and girls needed accommodation in curriculum planning. Notions of girls as future homemakers underpinned the Committee's argument that Victorian pioneers "had sacrificed the individuality of womanhood on the austere altar of sex equality," rather than educating girls and boys differently in line with their future roles. The Committee emphasized the importance of physiological and emotional differences between the sexes. Arguments about "overstrain" and the need to protect girls from competition surfaced alongside imperialist discourse that urged women to preserve their health as a moral duty to empire and "race."[47]

Headmistresses argued that music, art, and (compulsory) domestic subjects, which were girls' strengths, should count for girls toward the School (leaving) Certificate,[48] which provided access to higher education and the professions. The 1919 Sex Disqualification Removal Act made discrimination in entry to the professions illegal on the grounds of sex, apart from entry to the Church and higher branches of the civil service. Alongside academic preparation for School Certificate, girls' schools continued to emphasize the caring role and notions of "service" expected of women in society. Interwar debate about coeducation hinged around sensitive girls "civilizing" boys and girls becoming more assertive, rather than increasing educational attainment. Debate was not concerned to increase the number of girls' secondary schools, despite the fact that the smaller numbers of schools meant that girls had to perform better than boys in intelligence tests to gain a grammar school place. On the eve of the Second World War, 226,614 girls and 247,389 boys were attending 446 girls', 513 boys', and 513 mixed state-maintained secondary schools in England and Wales.[49]

Girls' Secondary Education Post 1944

The Post-1944 Education Act (England) provided all children up to the age of fifteen with the right to free secondary education. The Act did not prescribe a tripartite system, but the majority of children were allocated via the 11+ examination to grammar, technical, or secondary modern schools. Girls continued to outperform boys in the 11+ examination and results were weighted so that fewer girls than the results merited progressed to the grammar schools that gave access to higher education and the professions. Technical schools were too few to be available for many girls. Secondary modern schools provided education for the majority, and girls were likely to leave at fifteen without any qualifications. The girls' secondary modern curriculum was humanities-based and heavily domesticated, with little access to science and technical craft subjects. The Norwood (1943), Crowther (1959), and Newsom (1963) Reports all emphasized girls' future destiny in the family. Domestic subjects were linked with a postwar focus on the mother and family, and continued to be targeted at the less able and the lower social classes in the low-status secondary modern schools; they were neglected in the grammar schools.[50]

In *The Education of Girls* (1948), John Newsom (1910–1971), subsequently author of the government *Report on the Education of Children of Average and Lower than Average Ability* (1963), criticized the girls' grammar schools for overemphasizing public examinations and professional careers. He portrayed clever girls neglecting domestic subjects, and insisted that girls had different needs and needed preparation for homemaking and motherhood. The 1944 Act had removed the marriage bar, prevalent for women teachers since the 1920s and 1930s, and Newsom recommended the appointment of a mix of unmarried and happily married teachers, who were to look attractive to give the girls the impression that they were not spurning marriage. The Board of Education continued to oppose coeducation, believing it would have a detrimental effect on boys. Women teachers also argued that mixed schooling would be dominated by male teachers with less concern for the education of girls.[51]

By the 1950s, the attachment theory of the child psychiatrist John Bowlby (1910–1971) was highly influential. This saw a mother's presence as essential to the care of babies and of children beyond school age, and even the most highly educated women were expected to stay at home with their children. Woman's dual role of paid work and family was reflected in the Crowther Report (1959) on the education of 15–18-year-olds.[52] At the start of the 1960s the majority of middle-class girls attended single-sex grammar schools, modelled on the independent sector and oriented toward higher education, while more than half of the girls from the working class attended mixed secondary modern schools. Many girls were significantly ahead of boys at various points in their school career but continued to leave at an earlier age and to underachieve in external examinations. In Scotland, where the 1936 (Scotland) Act had brought about secondary education for all, only 54 percent of girls completed any kind of five-year course, as opposed to 60 percent of boys.[53]

With the growth of social democratic ideology during the 1960s, single-sex education in grammar schools came to be seen as elitist. The shift toward comprehensive education in England led to closure and amalgamations for many single-sex schools. In Ireland, in contrast, single-sex secondary schools remained more common.[54] During the 1960s and 1970s, there was virtually no debate about the impact of comprehensive schools on girls' education in England. The exception was a three-volume study that focussed largely on the social advantages of coeducation.[55] While some middle-class girls benefited from the introduction of comprehensive education and the raising of the school-leaving age, the benefits for girls were uneven. Feminist educationists were highly critical of comprehensives for gendered curriculum patterns that left many working-class girls without qualifications, the disappearance of women from senior teaching posts, and the masculinist culture of many schools. The Women's Liberation Movement of the 1960s and 1970s called for equal educational opportunities for both sexes, and the Sex Discrimination Act (1975) aimed to end discrimination in entry to mixed schools and direct discrimination in the type of courses offered to pupils, although it did not address gendered issues within the curriculum.[56]

The New Right Thatcherite policies of the 1970s and 1980s, which focussed on selective education, the curriculum, and parental choice, proved contradictory for girls. The secondary school curriculum was standardized in the Education Reform Act (1988), which aimed to meet the needs of the economy and industry in line with market ideologies, rather than to advance equal opportunities. The result was an entitlement curriculum that made compulsory those areas where girls were underrepresented and provided the context for initiatives around science for girls. As Arnot et al. argue, downgrading the female domestic courses could be seen to release girls from domestic ideologies, or, alternatively, demonstrate the Conservative government's lack of concern for those subjects in which girls excelled.[57]

Despite the increase in immigration of families from the Caribbean and Asia from the 1950s onward, black and Asian girls were largely invisible in educational theory and policy or were portrayed through negative stereotypes. Asian girls were the highest achieving among the working-class sample in a study conducted in 1985. African Caribbean girls were also more likely than boys to stay on at school to gain qualifications equivalent to those of their white peers. In the 1980s such evidence was used to suggest (inappropriately) that "dominant" African Caribbean girls and women were the cause of male disadvantage and underachievement. Official reports, like the Rampton Report on West Indians in Britain (1981) and the Swann Report (1985), continued to define the "problem" for West Indians in Britain as underachievement and to see gender as insignificant. The 1990s educational

statistics, in contrast, illustrated that black girls stayed on longer at school than
white working-class children and were better qualified than black males.[58]

Colonial Education

The complexities of continuity and change around girls' education and ethnicity
are further apparent when British colonial and settler societies are considered. Tim
Allender shows that in India, where the British Raj intersected with a long-standing
but highly regionalized heritage of women's education, the British State became
adept at suddenly embracing and then abandoning various girls' school initiatives;
and training female teachers, rather than educating schoolgirls, became a preoccupation.[59] British women reformers and educationists largely represented the Indian
women as their "Other," passively awaiting white women's help. The zenana (female
living quarters) formed a symbol of women's oppression in feminist, missionary, and
educational discourse. The image of the "needy" Indian women encouraged British
women living in England to think of themselves as committed to an imperial "civilizing" mission in which they assumed responsibility for the Indian woman's "enlightenment" and "rescue." This image was at variance with the work for girls' education by
Indian women, such as Pandita Ramabai, Rokeya Sakhawat Hossein, and Cornelia
Sorabji, who combined aspects of British and indigenous educational patterns.[60]

Education for indigenous populations in many British colonies and settler societies formed part of a "civilizing" mission that actively discouraged indigenous language, belief systems, and culture. Indigenous girls were to be transformed into
models of Victorian domestic femininity as homemakers, wives, and mothers, and
practices such as polygamy were discouraged. Early missionary teachers in New
Zealand focussed on Maori girls before Pakeha (white) girls and aimed to transform Maori into the guardians of morality and Christianity. As wives and mothers, they were to take responsibility for the physical and moral health of the future
generation in a model that linked motherhood and the nation. Maori were to be
"civilized" in the denominational boarding schools through a Victorian girls' curriculum based on the English grammar school and by engaging in the domestic
tasks that supported the daily operation of the school. At Hukarere, the Anglican
native school for girls subsidized by the State, girls received a classical curriculum of
English, Latin, algebra, physiology, drawing, history, drill, singing, and dressmaking. Taught by well-qualified staff, some Maori girls progressed to higher education
and the professions.[61]

In societies like Australia, early nineteenth-century governesses from Britain set up
ladies' academies to teach "accomplishments" to daughters of white settlers. Many of
the new academic girls' schools had links with the nineteenth-century English wom-
en's movement, while convent schools taught forms of domesticity and religion that
drew upon Catholic traditions. The Advanced School for Girls in Adelaide (1875), the
first secondary school to be established as part of a public school system in Australia,
provided an academic curriculum oriented toward the public examinations of the
local university and aimed to follow the new forms of English public schools for girls

created during the second half of the nineteenth century. Unlike the United States, the Australian public high school was selective and adopted many of the features of the reformed English grammar and public school traditions.[62]

Conclusion

Studies of girls' schooling in Britain and the British colonies demonstrate that secondary education advanced the position of women, while enacting aspects of social reproduction that related to local, national, and colonial specificities and international debates. Despite broader social and educational change for women and the fact that girls consistently outperform boys in examinations, studies continue to demonstrate the historical legacy of gendered educational landscapes and practice.[63] In 1997, Stephen Ball and Sharon Gerwitz noted that, like Francis Buss, Dorothea Beale, and the Girls Public Day School Trust in the nineteenth century, single-sex girls' state-maintained secondary schools in England stressed their "cloistered" ethos and fostered responsibility, quietness, femininity, and a particular role for the arts, dance, and music.[64] At the same time, however, they put forward a form of feminism concerned with having academic standards that equaled those of boys.[65]

Notes

1. Jane McDermid, "Women and Education," in *Women's History: Britain, 1850–1945*, ed. June Purvis (London: University College London Press, 1995), 125.
2. The term "public" school shifted from the narrow meaning attributed to the public schools recognized by the Clarendon Commission in 1861, through the broader meaning of the Headmasters Conference, to today's independent schools financed by fees paid by parents, rather than supported by rates and taxes. Richard Aldrich, *Public or Private Education? Lessons from History* (London: Woburn Press, 2003), 5.
3. Ruth Watts, "Gendering the Story: Change in the History of Education," *History of Education* 34, 3 (2005): 225–241.
4. Jane Rendall, "'Women that would plague me with rational conversation': Aspiring Women and Scottish Whigs, c1790–1830," in *Women, Gender and Enlightenment*, ed. Sarah Knott and Barbara Taylor (Basingstoke: Palgrave, 2005), 326–348.
5. Deirdre Raftery, Jane McDermid, and Gareth Elwyn Jones, "Social Change and Education in Ireland, Scotland and Wales: Historiography on Nineteenth-century Schooling," *History of Education* 36, 4 (2007): 447–463.
6. Carol Dyhouse, *Girls Growing up in Late Victorian and Edwardian England* (London, Routledge and Kegan Paul, 1981); Joyce Senders Pedersen, *The Reform of Girls' Secondary and Higher Education in Victorian England: A Study of Elites and Educational Change* (New York and London: Garland Publishing Inc., 1987).
7. Maria Tamboukou, *Women, Education and the Self: A Foucauldian Perspective* (Basingstoke: Palgrave, 2003), introduction.
8. Joyce Goodman and Sylvia Harrop, eds., *Women Educational-Policy Making and Administration in England: Authoritative Women since 1800* (London: Routledge, 2000).
9. Monica Bolufer Peruga, "Gender and the Reasoning Mind," in Knott and Taylor, *Women, Gender and Enlightenment*, 189.

10. Ruth Perry, "Astell, Mary (1666–1731)," in *Oxford Dictionary of National Biography* (Oxford: Oxford University Press, 2004), online ed., October 2007 [http://www.oxforddnb.com/view/article/814, accessed, May 30, 2009].

11. Jane Rendall, "Women and the Enlightenment c1690–1800," in *Women's History: Britain, 1700–1850: An Introduction*, ed. Hannah Barker and Elaine Chalus (London: University College London Press, 2005), 9–32.

12. Michèle Cohen, "Gender and 'Method' in Eighteenth-century English Education," *History of Education* 33, 5 (2004): 587; and "'A Little Learning'? The Curriculum and the Construction of Gender Difference in the Long Eighteenth Century," *British Journal for Eighteenth-Century Studies* 29, 3 (2006): 323.

13. Susan Skedd, "Women Teachers and the Expansion of Girls' Schooling in England c1760–1820," in *Gender in Eighteenth-Century England*, ed. Hannah Barker and Elaine Chalus (London: University College London Press, 1997), 101–125; Christina de Bellaigue, *Educating Women: Schooling and Identity in England and France, 1800–1867* (Oxford: Oxford University Press, 2007); Carol Percy, "Learning and Virtue: English Grammar and the Eighteenth Century Girls' School," in *Educating the Child in Enlightenment Britain*, ed. Mary Hilton and Jill Shefrin (Surrey: Ashgate, 2009), 83–84; Bellaigue, *Educating Women*, 173, 174; Michèle Cohen, "Gender and the Public Private Debate," in Aldrich, *Public or Private*, 25; idem, "'A Little Learning?,'" 322.

14. Cohen, "Gender and Method," 585.

15. Debbie Simonton, "Women and Education," in Barker and Chalus, *Women's History in Britain*, 45; Anne Stott, "Evangelicalism and Enlightenment: The Educational Agenda of Hannah More," in Hilton and Shefrin, *Educating the Child*, 50.

16. Simonton, "Women and Education," 46; Jane Rendall, *The Origins of Modern Feminism: Women in Britain, France and the United States, 1780–1860* (Basingstoke: Macmillan, 1985), 63; idem, "Women and the Enlightenment," 25.

17. Camilla Leach, "Quaker Women and Education from the Late Eighteenth to the mid-Nineteenth Century," (Ph. D. Diss., University of Winchester, 2003), 91, 108.

18. Bellaigue, *Educating Women*, 14, 21, 74; Skedd, "Women Teachers."

19. Bellaigue, *Educating Women*, 46, 52.

20. Wendy Robinson, *Pupil-teachers and Their Professional Training in Pupil-teacher Centres in England and Wales, 1870–1914* (Lampeter: Edwin Mellen Press, 2003).

21. Elizabeth Edwards, *Women in Teacher Training Colleges, 1900–1960* (London: UCL Press, 2001), 16.

22. Carmen M. Mangion, *Contested Identities: Catholic Women Religious in Nineteenth-Century England and Wales* (Manchester: Manchester University Press, 2008), 142.

23. Barbara Walsh, *Roman Catholic Nuns in England and Wales, 1800–1937: A Social History* (Dublin: Irish Academic Press, 2002), 42.

24. Bellaigue, *Educating Women*, 54.

25. Andrea Jacobs, "'The girls have done very decidedly better than the boys'; Girls and Examinations 1860–1902," *Journal of Educational Administration and History* 33, 2 (2001); 120–136.

26. Bellaigue, *Educating Women*, 180.

27. Pedersen, *The Reform of Girls' Secondary Education*.

28. Cohen, "Language and Meaning," 80–87.

29. This paragraph and the next three draw heavily on Lindy Moore, "Young Ladies' Institutions: The Development of Secondary Schools for Girls in Scotland, 1833–1870," *History of Education* 32, 3 (2003): 249–272.

30. Bellaigue, *Educating Women*, 113, 171–172.

31. Dyhouse, *Girls Growing Up*, 56; Sheila Fletcher, *Feminists and Bureaucrats: A Study in the Development of Girls' Education in the Nineteenth Century* (Cambridge: Cambridge University Press, 1980), 192–217; Joyce Goodman, "Girls' Public Day School Company (*act.* 1872–1905)," in *Oxford Dictionary of National Biography*, online ed., Oxford University Press, Feb 2009, http://www.oxforddnb.com/view/theme/94164 (accessed June 6, 2009).

32. Julia Bush, *Women Against the Vote: Female Anti-Suffragism in Britain* (Oxford: Oxford University Press, 2007).

33. Dyhouse, *Girls Growing Up*, 59, 162; Kathleen McCrone, "Playing the Game and Playing the Piano: Physical Culture and Culture at Girls' Public Schools c1850–1914," in *The Private Schooling of Girls Past and Present*, ed. Geoffrey Walford (London: Woburn Press, 1993), 33; Felicity Hunt, "Divided Aims: the Educational Implications of Opposing Ideologies in Girls' Secondary Schooling 1850–1940, in *Lessons for Life. The Schooling of Girls and Women, 1850–1950*, ed. Felicity Hunt (Oxford: Blackwell, 1987), 11; idem, *Gender and Policy in English Education, 1902–1944* (Brighton: Harvester Wheatsheaf, 1991), 26.

34. This paragraph and the next two rely heavily on Gareth Elwyn Jones, *Education and Female Emancipation: The Welsh Experience, 1847–1914* (Cardiff: University of Wales Press, 1990), 35–40, 114, 116, 117, 134, 159, 160, 167.

35. This and the following paragraph rely heavily on Deirdre Raftery and Susan M Parkes, *Female Education in Ireland 1700–1900: Minerva or Madonna* (Dublin: Irish Academic Press, 2007), 68, 70, 73; Deirdre Raftery, "The Higher Education of Women in Ireland, 1860–1904," in *A Danger to the Men: A History of Women in Trinity College Dublin, 1904–2004*, ed. Susan Parkes (Dublin: The Lilliput Press, 2004), 9–13.

36. Mangion, *Contested Identities*, 143, 145; Walsh, *Roman Catholic Nuns*, 5, 41, 57; Peter Doyle, *Mitres and Missions in Lancashire: The Roman Catholic Diocese of Liverpool, 1850–2000* (Liverpool: The Bluecoat Press, 2005), 136.

37. Robinson, *Pupil-teachers*, 67.

38. Meriel Vlaeminke, *The English Higher Grade Schools. A Lost Opportunity* (London: Woburn Press, 2000), 40, 130–231.

39. Jane McDermid, "'Make way for the lass of parts?' Class and Gender in the Secondary Schooling of Girls in late Nineteenth Century Glasgow." http://www.inrp.fr/she/ische/abstracts/mcdermid.pdf (accessed May 25, 2009).

40. Robinson, *Pupil-teachers*, 12, 34.

41. Dina Copelman, *London's Women Teachers: Gender, Class and Feminism, 1870–1930* (London: Routledge, 1996), 135.

42. Gareth Elwyn Jones, *Controls and Conflicts in Welsh Secondary Education, 1889–1944* (Cardiff: University of Wales Press, 1982), 7, 9.

43. Lindsay Paterson, *Scottish Education in the Twentieth Century* (Edinburgh: Edinburgh University Press, 2003), 53; McDermid "Women and Education," 122.

44. Hunt, "Social Class," 34, 36; and "Divided Aims," 9, 10, 12, 13; Dyhouse, *Girls Growing Up*, 162, 166, 157, 162, 166.

45. Dyhouse, *Girls Growing Up*, 123–125; Hunt, *Gender and Policy*, 29, 31–32.

46. Dyhouse, *Girls Growing Up*, 131–132.

47. Hunt, *Gender and Policy*, 38; idem, "Divided Aims," 7; Dyhouse, *Girls Growing Up*, 132–136.

48. Hunt, "Divided Aims," 19, 20.

49. Penny Summerfield, "Cultural Reproduction in the Education of Girls: a Study of Girls' Secondary Schooling in Two Lancashire Towns, 1900–1950," in Hunt, *Lessons for Life*, 154; Kevin Brehony, "Co-education: Perspectives and Debates in the Early Twentieth Century," in *Coeducation Reconsidered*, ed. Rosemary Deem (Milton Keynes: Open University Press, 1984), 1; Hunt, "Divided Aims," 20.

50. Stephanie Spencer, *Gender, Work and Education in Britain in the 1950s* (Basingstoke: Palgrave, 2005), 51; Madeleine Arnot, Miriam David and Gaby Wiener, *Closing the Gender Gap: Postwar Education and Social Change* (London: Polity Press, 1999), 39, 113; Rosemary Deem, "State Policy and Ideology in the Education of Women 1944–1980," in *Equality and Inequality in Education Policy*, ed. Liz Dawtrey, Janet Holland, and Meriel Hammer (Clevedon: Multilingual Matters, 1995), 35; McDermid, "Women and Education," 124.

51. McDermid, "Women and Education," 124; Denis Dean, "Education for Moral Improvement, Domesticity and Social Cohesion: the Labour Government, 1945–1951," in Dawtrey et al, *Equality and Inequality*, 24, 26.

52. Arnot et al., *Closing the Gender Gap*, 40; Deem, "State Policy," 36; Spencer, *Gender, Work and Education*, 50.

53. Arnot et al., *Closing the Gender Gap*, 69; Roy Lowe, *Schooling and Social Change, 1964–1990* (London: Routledge, 1997), 100; Paterson, *Scottish Education*, 129, 131; Alan Smithers

and Pamela Robinson, *Co-educational and Single-Sex Schooling* (Manchester: Centre for Education and Employment Research, 1995), 5.

54. Kathleen Lynch, "Republic of Ireland," in *Girls and Young Women in Education: A European Perspective*, ed. Maggie Wilson (Oxford: Pergamon, 1991), 115–136, see 120.

55. Reginald Rowland Dale, *Mixed or Single Sex School?* vol. 1 (London: Routledge, 1969); *Mixed or Single Sex School? Some Social Aspects.* vol. 2 (London: Routledge, 1971); and *Mixed or Single Sex School? Attainment, Attitudes and Overview.* vol. 3 (London: Routledge, 1974).

56. Madeleine Arnot, "Feminism, Education and the New Right," in Dawtrey et al, *Equality and Inequality*, 166; Heidi Mirza, "The Myth of Underachievement," in Dawtrey et al, *Equality and Inequality*, 183, 186; Arnot et al., *Closing the Gender Gap*, 40, 69, 113; Deem, "State Policy," 38, 40.

57. Deem, "State Policy," 42; Sheila Miles and Chris Middleton, "Girls Education in the Balance: the Era and Inequality," in Dawtrey et al, *Equality and Inequality*, 124, 125; Arnot, "Feminism, Education," 175.

58. Arnot et al., *Closing the Gender Gap*, 115; Mirza, "The Myth of Underachievement," 186, 189, 195.

59. Tim Allender, "Imagining Innovation: State Agendas for Women's Education in Colonial India," *History of Education Researcher* 80 (November 2007): 100–112.

60. Antoinette Burton, *Burdens of History: British Feminists, Indian Women and Imperial Culture, 1865–1915* (Chapel Hill, NC: University of North Carolina Press, 1994); Barnita Bagchi, "Two Lives: Voices, Resources and Networks in the History of Female Education in Bengal and South Asia," *Women's History Review*, 19, 1 (2010): 51–69.

61. Kay Morris Matthews, *In Their Own Right. Women and Higher Education in New Zealand before 1945* (Wellington: NZCER Press, 2008), 36–37, 49.

62. Craig Campbell and Geoffrey Sherrington, *The Comprehensive Public High School. Historical Perspectives* (New York: Palgrave, 2006), 17, 19–21.

63. Joyce Goodman, "A Cloistered Ethos? Landscapes of Learning and English Secondary Schools for Girls: An Historical Perspective," *Paedagogica Historica* 41, 4/5 (2005): 589–603.

64. Stephen Ball and Sharon Gewirtz, "Girls in the Education Market: Choice, Competition and Complexity," *Gender and Education* 9, 2 (1997), 210.

65. Diana Leonard, "The Debate About Co-education," in *Targeting Underachievement: Boys or Girls*, ed. S. Kemal, D. Leonard, M. Pringle, and S. Sadeque (London: Institute of Education, 1995), 25.

Chapter 2

Culture and Catholicism: France

Rebecca Rogers

France has long been recognized as a country where cultivated women pen elegant letters, write romantic novels, or converse wittily in a salon. This tradition of the cultured woman has produced the likes of the scientist Emilie de Châtelet in the eighteenth century, the novelist George Sand in the nineteenth century, or the philosopher Simone de Beauvoir in the twentieth century.[1] That women could achieve fame and recognition through their learning and knowledge suggests a special relationship to education, culture, and femininity, which is by now well researched.

In the late 1960s, the new social history tackled the history of education via the Church-State conflict and literacy. While girls were not absent from the narrative that unfolded, they were mainly presented as the objects of a struggle between religious and secular forces. In this narrative, serious education for girls really only emerged in the late nineteenth century when the State created a public secondary system for girls that echoed—with a difference—what Napoleon Bonaparte had set up for boys in 1802. This chapter builds on existing scholarship to argue that this vision of "serious education" sorely underestimates earlier achievements, while rendering invisible thousands of religious and lay women teachers, who opened schools, wrote textbooks, and sought to promote a certain model of the cultured French woman who, while not a *femme savante*, could nonetheless talk about history and literature, converse in foreign languages, and play the piano.

The concern to analyze the history of girls' education emerged early in France with the pioneering work of Françoise Mayeur on the creation of girls' secondary *collèges* and *lycées* (Camille Sée law of December 1880).[2] Followed swiftly by a more general book about middle-class girls' education in the nineteenth century, her work highlighted the importance of the early Third Republic educational reforms that not only established a public secondary system for girls but also normal schools to train future schoolteachers.[3] Her detailed social analysis of both students and teachers illustrated the middle-class nature of these institutions as well as the evolution in the twentieth century toward increasing similarity with boys' education as girls aspired to pass the baccalaureate exam initially reserved for boys. Her study of the private sector before the 1880s suggested, however, that secondary education existed well before the Republican laws but its characteristics were harder to discern and more fragmented. Above all, she belied what many histories of education suggested: that religious institutions were mere agents of indoctrination, under the influence of a male clergy.

My own work, as well as that of Christina de Bellaigue, has pursued the exploration of the nineteenth century and particularly the characteristics of an active private sector.[4] The Catholic model of the boarding school for bourgeois girls prevailed in this sector and durably marked the nature of the education offered. More than my predecessor, I have sought to emphasize the way women teachers took advantage of periods of flux—after the Revolution or in 1848—to state the need to broaden women's minds and how attention to their initiatives changes our understanding of French women's history and attenuates the triumphalism of the Republican record.[5] Further work remains to be done, however, on the specific pedagogical practices and the clientele of religious teaching orders and on the eighteenth-century legacy of girls' schooling. To date, Martine Sonnet's fine study of girls' schooling in early modern Paris has attracted few imitations.[6] As a result, while we know that elite boarding schools existed throughout France, few social histories question the dominant narrative for the eighteenth century of educational decline in aristocratic circles, based mainly on hearsay and a few student memoirs about the decadence of Louis XIV's famous *Maison Royale de Saint-Cyr*. Similarly, the history of secondary girls' education in the twentieth century awaits reappraisal. The private sector never disappeared in girls' secondary education, but most works focus on public institutions and their women teachers, and very few studies move into the postwar period to explore the changes provoked by the move toward coeducation and the democratization of the secondary level of schooling.[7]

What follows then is a succinct effort to show the characteristics of what constituted post-elementary nonvocational schooling for girls in France over the past three hundred years. Without overstating the weight of political chronology in this narrative, three broad periods can be distinguished: the eighteenth century up to and including the French Revolution; the transformations of the nineteenth century until 1880; and then the period from the Republican educational laws until the final quarter of the twentieth century when most secondary girls' schools disappeared in favor of coeducation.

Eighteenth-Century Schools, Debates, and Revolutionary Intentions

In the eighteenth century, the expression "secondary education" was not much used to designate either boys' or girls' education. In this Old Regime hierarchical society, most children received very little education at all, and girls received less than boys. What can be considered secondary education for girls involved lessons that went beyond religion, reading, writing, and arithmetic to include literature, history, geography, foreign languages, and the arts. Certain aristocratic families hired governesses or tutors for their daughters, but there were also schools for wealthy girls, run most frequently by religious orders. While these schools were increasingly the object of criticism, their existence testifies to a long tradition of offering postprimary education to girls in order for them to acquire not only the polish but also the learning associated with upper-class lifestyles.

Institutions

The most illustrious of girls' institutions was unquestionably the *Maison Royale de Saint-Cyr* founded in 1686 to educate the daughters of Louis XIVth's nobility. Inspired by Fénelon, the author of an oft-cited treatise *L'éducation des filles* (1687), Mme de Maintenon established this school with the intention of reforming the French nobility through the education of noble girls.[8] The school set up four different classes where students progressed from the study of the rudiments and the catechism to that of history, geography, music, language, drawing, dance, and the indispensable moral education. The school library testifies to an education that, while socially conservative, nonetheless introduced students to serious reading matter, notably in history. Girls were admitted between the ages of seven and twelve and stayed until twenty. Their cloistered existence resembled that of nuns, and indeed many students went on to become nuns.

Other institutions existed as well, often the product of Counter-Reformation efforts to provide serious moral schooling for both boys and girls. The most prominent of such institutions were those run by the Ursulines, the Congrégation Notre-Dame and the Visitandines: they all focused their efforts on boarding-school students and were located in cities, catering to relatively privileged urban families. The women who taught in these institutions themselves came from such families.[9] In Paris, a relatively wide range of boarding schools catered to the bourgeoisie and the aristocracy, offering instruction that went beyond the rudiments even for the relatively young. The daughter of a Parisian engraver, Manon Philipon, for example, spent a year at the age of eleven in the school run by the Congrégation Notre-Dame to prepare for her first Communion. Lessons in grammar, history, and geography, as well as music and drawing, accompanied those in religion. This brief year of school learning was reinforced, however, within her home environment where she read in the family library and received lessons, including in Latin, from individual tutors. This commitment to educating a young girl was probably fairly unusual, however, and relatively few families could afford the boarding costs involved in sending their daughters to school.[10] Overall, historians have highlighted the generally mediocre quality of girls' education, echoing, in this respect, contemporary opinion.

Philosophers and Pedagogues

Given the importance attached to culture and knowledge in France, the issue of women's access to this culture inevitably figured in Enlightenment debates. As early as 1673, François Poulain de la Barre had published a treatise in which he argued that "the mind had no sex" and thus women should have access to the same studies as boys, be able to enter the University, and acquire professional degrees. His strikingly modern vision of gender equality remained relatively isolated during the eighteenth century, as most of the famous (male) philosophers adhered to the idea that biological differences did indeed influence "moral" differences between the sexes. As a result, while girls could receive some education, few argued that this education should be identical to that of boys. Most famously, the Genevan philosopher Jean-Jacques Rousseau argued in his pedagogical novel *Emile* (1762): "Thus the whole education of women ought to be relative to men. To please them, to be useful to them, to make themselves loved and honored by them, to educate them when young."[11] Motherhood in this vision of gender relationships justified the development of girls' education.

Many pedagogues took to their pens to propose educational models where girls could acquire the knowledge necessary to make them worthy wives and mothers and in the process advocated more rigorous education. In 1730 the abbé de Saint-Pierre argued for the need to create girls' *collèges* (secondary schools), similar to those that existed for boys. More commonly, cultivated men and women published tracts and treatises that criticized the limited nature of girls' education and offered suggestions to improve it. In 1728 the salonière Mme de Lambert (1647–1733) published her *Avis d'une mère* (Suggestions of a mother) in which she criticized the frivolous character of girls' education and advocated moral training; sixty years later Mme de Miremont (1735–1811) saw fit to print seven volumes of her *Traité de l'éducation féminine* (Treatise of Feminine Education) that focussed a great deal on teacher training.[12] Just as rising literacy rates allowed more girls access to the written word—literacy rates for women increased from 14 to 27 percent, while men's literacy increased from 29 to 47 percent—, the print revolution opened the way for educated women to spread their views. These texts were not best-sellers, nor did they have an obvious impact on the nature of available schooling, still they testify to a concern among certain portions of the population to improve girls' education, a concern that would have more concrete results in the following century.

Teaching and pedagogy allowed some women to achieve genuine fame at the end of the eighteenth century. This was true, in particular, for Mme de Genlis, known as the educator of Princes since she taught the future King of France, Louis Philippe; her moral and didactic tales were widely read, commented, and translated in the Revolutionary period and well into the early decades of the nineteenth century.[13]

Revolutionary Contradictions

The intellectual and political ferment of the French Revolution encouraged an active debate about educational matters, but produced no blueprint for the future. More concretely, anticlerical measures thoroughly disrupted existing schools for girls, as members of religious orders scrambled to save their lives. As a result, the revolutionary decade spawned an outpouring of often contradictory educational proposals with little concrete effect. Still, these proposals reveal the extent to which girls' education was not a priority for most revolutionaries. The creation of the *École normale de l'an III* (1795) or the scientifically oriented secondary schools, the *écoles centrales*, were masculine initiatives despite pressure from women calling for equal opportunities and equal treatment with men. Some argued that girls should receive an education that would allow them to enter the professions and be economically independent, while others adopted a more feminist tone accusing men of deliberately restricting women's access to reason: "It's reason that you fear," wrote an anonymous woman in *Le Courrier de l'Hymen, Journal des Dames* in 1791.[14] The liberal aristocratic philosopher Condorcet was the most prominent intellectual of the period to propose that boys and girls receive the same education within the same institutions. Although he did not state explicitly that girls should attend the secondary schools he envisioned, this was certainly the tenor of his project. Like other proposals of the period, however, it never saw the light of day, and when one turns to institutional realities, the revolution did more to destroy existing girls' schools than to establish new ones in their place.

One of the few institutions to emerge successfully from these years of political turmoil was run by Jeanne Henriette Campan (1752–1822) in the Parisian suburb of Saint-Germain-en-Laye. As lady-in-waiting to Marie-Antoinette, Campan narrowly escaped execution during the Revolution. An astute reader of political currents, she used her acquired education as well as her social connections to found a boarding school in 1795. Within a year her school had one hundred students and offered wealthy girls an accomplishment-oriented curriculum that nonetheless involved hours of more intellectual instruction.[15] Visiting foreign dignitaries, such as the Englishwoman Fanny d'Arblay, attended the lavish prize-giving ceremonies and commented not only on the content of the girls' instruction but also on the high percentage of "the rich and the gay."[16] So while the Revolution passed no significant legislation to promote girls' education, it nonetheless offered opportunities for a few. Unquestionably, Mme Campan was the woman who benefited the most from the institutional tabula rasa and her notoriety only increased in the new century as France moved from a Republic to an Empire under Napoleon.

Warring Factions? Private Institutions and the State (1800–1880)

The rebuilding of the social order following the Revolution created opportunities for teachers and pedagogues who used arguments about women's roles as mothers and educators to defend more serious studies for girls.[17] Although the distinction between girls' primary and secondary education was not as clearly defined as for boys, most contemporaries associated the latter with costly boarding schools and the middle classes: in professional trade books secondary institutions were listed as *pensionnats de demoiselles* (boarding schools for young ladies). These schools often included classes for young students learning the rudiments, while older girls had lessons in literature, history, geography, and foreign languages, although not Latin and Greek. Both men and women taught within these schools, but only women directed them. While lay schools were more numerous in cities than religious schools, the latter tended to take in more students and increasingly were perceived as dominating the educational market for girls. Until the acrimonious debates between clericals and anticlericals in the second half of the century, however, both lay and religious institutions existed relatively harmoniously side by side, offering a very similar curriculum to families who could afford it.

Early Nineteenth-Century Models in Girls' Schooling

While no feminist, Napoleon Bonaparte established in 1805 the schools of the Legion of Honor, which became a model for feminine secondary school education throughout the century.[18] Although these were not, strictly speaking, public institutions, the State paid the teachers' salaries and provided fellowships for the students who then spent between five and six years in a large boarding-school setting remarkably similar in ethos to that of the Napoleonic *lycée*, notably with its emphasis on discipline and emulation. The Emperor founded a total of five schools, three of which survived his fall from power in 1815; his concern was to succor the families of his soldiers who had died on the field of battle or who had distinguished themselves, thus earning the Legion of Honor. Through the education of their daughters he meant to cement their commitment to the French nation as well as to his person. The education he envisioned was fairly limited:

> What will be taught to the *demoiselles* who will be brought up at Ecouen? We must begin with religion in all its rigor; in this matter do not admit any modification. In a public institution for demoiselles, religion is a serious matter; whatever else may be said about it, it is the surest guarantee for mothers and for husbands. Make believers of them, not reasoners.[19]

Jeanne Campan, the headmistress of the first school at Ecouen, had very different ideas, however, and these ultimately won out within the classroom, just as her pedagogical writings spread far beyond the borders of France. Her students studied literature, ancient, sacred, and French history, geography and geometry, some natural sciences (especially botany), in addition to more feminine lessons in sewing, hygiene, and cooking. This plan of study differed from the better Old Regime convent schools in the time allotted for literature, geometry, and the sciences. Campan's private writings reveal, moreover, that she envisioned her school in Ecouen as a "University of women," and she had ambitious plans to position it at the center of a vast network of girls' schools in France. Although her own influence faded with the fall of the Empire, many students of the Legion of Honor went on to active careers, particularly in teaching, rather than devoting themselves to home life. Schools modelled on her institution were opened in Germany and Italy, while in France Legion of Honor course books became a staple in an increasingly active schoolbook market, spreading an ideal of serious secondary education for girls.

The Napoleonic years were also marked by the reemergence of female religious congregations, following their disappearance during the Revolution. Old Regime orders, such as the Ursulines, reopened schools, and new congregations were created, including quite a few that specialized in boarding-school education for the middle and upper classes. In the postrevolutionary period, the Ladies of the Sacred Heart were among the most influential and elitist of these new religious teachers, whose primary mission was to rechristianize French society through the education of girls.

By 1864, they ran forty-four boarding schools in France and another forty-two had opened in foreign or colonial lands, offering an educational program that resembled the secondary program of study within the Legion of Honor schools.[20] The quality of education within these schools varied greatly, however, prompting the liberal aristocrat Eugénie Milleret's decision to found the Dames de l'Assomption in 1839. Critical of superficial, excessively religious, and accomplishment-oriented curricula, she promoted serious learning for girls, in particular through the study of Latin and theology that were traditionally reserved for boys. Milleret's institution demonstrated that the commitment to serious learning existed within religious schools, although not for vocational purposes.

By the 1850s, private secondary schools for girls were spread throughout urban France. Certainly religious schools were more visible since they were installed in larger buildings and welcomed more students, but in trade books, individual lay teachers were more numerous, often setting up their school in an ordinary apartment building. Despite the lay status of teachers, religious lessons were always present, and Protestant and Jewish teachers opened schools as well in areas with significant religious minorities. In Paris, Mme Neymark and then Mme Kahn ran a school that attracted Jewish families from the early 1840s into the 1880s. Finally, one also finds schools aimed at adolescent British girls who would cross the Channel for a year or two in a boarding-school environment, where they learned French, as well as the arts.[21] Diversity characterized this period before the State's intervention.

The Politics of Girls' Educational Reform (1848–1880)

The development of opportunities to pursue postprimary education did not occur in an administrative vacuum despite the absence of state-supported institutions. As early as 1810 in Paris, municipal decrees established guidelines for the opening of schools, and in 1821 a ruling determined a hierarchy of secondary educational institutions as well as an examination system for the teachers and inspections for the schools. After the Guizot law of 1833 establishing primary education for boys, the capital city once again produced a ruling in 1837 that prefects were encouraged to apply throughout France.[22] These attempts to define what constituted secondary education occurred at a moment when gender relations were under debate within Saint-Simonian circles. This progressive socialist movement attracted women schoolteachers and women of letters who contributed their voices to a broader debate calling for better, more serious education for women. Between 1845 and 1848, liberal pedagogues published the *Revue de l'enseignement des femmes* (Review of Women's Education), the first journal specifically concerned with girls' education.

The reform movement reached a crescendo around the Revolution of 1848 and the proclamation of the Second French Republic. At the forefront of the educational battles, the Saint-Simonian headmistress Joséphine Bachellery penned a series of proposals that she sent to the Minister of Public Instruction, Hippolyte Carnot, calling for the creation of girls' *collèges*, as well as a superior normal school to train future secondary school teachers. This proposal followed earlier efforts by fellow Saint-Simonian Louise Dauriat to rid girls' schooling of male teachers, arguing they were monopolizing positions that should be held by women. Bachellery's militancy during these years took on a distinctly anticlerical edge, as she criticized the advantages teaching nuns held over lay women in the educational market, notably the fact they did not need a diploma to open a school. Her campaign to make teaching a lay woman's domain echoed a new concern to train women for more qualified professions. Yet another Saint-Simonian activist, Elisa Lemonnier, set in place the first vocational courses for girls in this period, an initiative that grew in significance in the 1860s. In the revolutionary backlash that brought Louis Napoleon to power as President in December 1848, all of the more ambitious plans for women's education fell to the wayside, except those concerning vocational schooling.

Like elsewhere in Europe, the 1860s constituted an important turning point for the development of girls' secondary education, thanks to the efforts of Victor Duruy, Napoleon III's liberal reforming Minister of Public Instruction (1863–1869). In addition to instituting reforms within the boys' secondary system, he developed the network of girls' primary schools and urged the creation of secondary courses for girls in 1867. His 1864 survey of girls' boarding schools had highlighted the increasing weight of religious orders who ran 2,338 out of 3,480 schools.[23] Although the survey demonstrated the existence of girls' schools throughout France, he promoted his call for reform by declaring: "One must found secondary education for girls, which in point of fact does not exist in France." He followed this highly political pronouncement by proposing the creation of a three-to-four-year course of study that proposed "a combination of general literary instruction, the study of foreign languages, and drawing with practical demonstrations of scientific knowledge." Most radically for the period, he eliminated religious instruction, arguing this was best accomplished within the home. While these courses were a far cry from the existing boys' *lycées* and received no direct State funding, they represented a significant step in the State's involvement in girls' secondary education. The teachers came from the local boys' *collèges* and *lycées* and classes were held generally in the local town hall. It was precisely this use of male professors and the absence of religious instruction that contributed to their downfall. The bishop of Orléans Félix Dupanloup led the opposition arguing that girls' education should remain in the hands of women, but he also denounced the vision of womanhood supposedly being promoted: "You want women freethinkers, unbelievers, more than that, women doctors of impropriety, professors of atheism, a type of unknown woman who would be dreadful."[24] In the short-run this campaign was successful, and most of the approximately sixty courses opened between 1867 and 1870 were short-lived and successful only among Protestants, Jews, and freethinkers. Still the courses remain important because they highlight the moment when girls' secondary education became an explicitly political issue and one that French republicans were increasingly ready to go to battle for.

These years prior to the French educational laws of the 1880s also saw the expansion of opportunities for girls' secondary schooling within the colonies. For the most part these efforts targeted middle-class European families, rather than local populations whose educational needs were deemed more rudimentary. Overwhelmingly, it was religious teaching orders who spread the messages of French "civilization" to the colonies, often aided in their task by public monies. Interestingly, the increasing concern among French anticlericals to pull women from the arms of the Church through an education shorn of religion and religious teachers had little impact in the colonies where religious orders were far more equipped to set up boarding schools.

The action of religious orders in the colonies began in the early modern period. The Ursulines were already present in Martinique as early as 1682. More concerted efforts to develop educational opportunities for girls occurred, however, in the nineteenth century, notably through the efforts of the first female missionary order, the Congrégation de Saint-Joseph de Cluny, founded in 1807 by Anne-Marie Javouhey. Rule-books and prize-giving ceremonies clearly indicate the way French models of girls' secondary education were transplanted to colonial soils, thanks to the presence of boarding schools for *demoiselles* both in the older colonies such as the Réunion, Guadaloupe, and Martinique, or the newer French possessions such as Pondicherry in India or New Caledonia. These institutions, which at times welcomed the daughters of local elites, were less present in Africa where missionary efforts targeted the African populations and offered mainly primary education with a strong vocational twist. In a settler colony such as Algeria, boarding schools for *demoiselles* were opened at the outset of colonization in the early 1830s; the Ladies of the Sacred Heart set up a boarding school in Algiers in 1843 that maintained the French tradition of strictly separating the student body by social class, while French lay women opened schools for European girls, including, specifically, for indigenous Jewish girls. While girls' secondary education was never a primary concern in French colonization, it is important to recognize the role of private initiative, and notably religious orders, in spreading a certain type of culture to the colonies.

The State Takes Over: From Single-Sex Schools to Coeducation (1880–1976)

The French State's investment in girls' secondary education began with a vengeance in the early years of the Third Republic (1870–1940). As Jules Ferry proclaimed in 1870: "Democracy must choose on pain of death; citizens must choose; women must belong to science or to the church." The educational laws between 1879 and 1886 that secularized both school programs and teaching personnel had greater consequences for girls than for boys and were part of a political program to win women over to Republican ideals. These measures unquestionably encouraged the development of girls' secondary education, but they should not obscure the existence of a network of such schools run both by lay and religious women as well as earlier feminist efforts to encourage such schooling.

The Republican Laws

On December 21, 1880, the Camille Sée law created a national and public system of *collèges* and *lycées* for girls. Although these schools took the name of their male counterparts, the content and organization of the girls' schools were distinctly feminine. Rather than a seven-year course of study leading to the prestigious baccalaureate degree that opened the door to university study, Sée's law set in place a three-year program, followed by an additional two years for the more intellectually

ambitious. The courses proposed were similar to those present in the better boarding schools: modern languages and French literary classics, history, and an introduction to geometry, physics, and natural history. Moral instruction replaced religion,
and in lieu of the baccalaureate, a *diplôme d'études secondaires* crowned the five years
of study in which the study of the arts, hygiene, domestic economy, and needlework
continued to proclaim the specificity of girls' cultural needs.

In order to ensure that studies were of a high quality, the Republican reformers
also created an *École normale supérieure* at Sèvres in the Parisian suburbs to train the
future women professors of the secondary system. This institution quickly acquired
a reputation as the French equivalent of a female college, where young women
studied for two years in order to pass the competitive examination, the *agrégation*,
which ensured a higher-paid position within the more demanding *lycées*. Above all,
the training at Sèvres developed an esprit de corps that did much toward forming
an ethos of female professionalism, as many memoirs testify.[25] Such an ethos also
emerged among primary schoolteachers after the Bert law of 1879 established lay
normal schools in all of the French departments. Clearly these years represented a
turning point.

The numbers of *collèges* and *lycées* grew slowly, limited by reliance on municipal
funding. In addition, the Sée law had not seen fit to create boarding schools at a
time when that sort of regimented living was under severe attack in boys' *lycées*; as a
result, the cost of opening a boarding school, to welcome students who did not live
in cities, fell upon individual municipal councils. Still the number of such schools
multiplied in France: in 1883 there were 10, this figure had risen to 103 by 1906. By
1914 some 35,000 girls received some form of public secondary education compared
to 100,000 boys.[26] The social groups involved in this education were solidly middle
class with a high percentage of Protestant students, although, of course, there were
regional variations. In general, however, the Catholic bourgeoisie failed to flock to
the new creations, preferring the more familiar environment of the religious boarding schools.

Indeed the continued success and growth of private secondary institutions for
girls has often been overlooked, as historians have focused on the new *lycées* and
collèges. The private sector thrived in the early twentieth century, and not just
among the socially conservative. In Paris, a number of institutions were founded
to prepare girls for the baccalaureate; they included the study of Latin and Greek,
absent from the state-run institutions. In addition, enterprising schoolmistresses
sought to provide better training for women secondary schoolteachers: Mathilde
Salomon opened professionally oriented classes in the collège Sévigné in 1885,
Mlle Desrez opened the *École normale catholique* in 1906, and two years later
Madeleine Daniélou created the *École normale libre* to prepare girls for university
examinations.

Increasingly, the feminine curriculum of the girls' *collèges* and *lycées* came under
criticism from families, teachers, professional associations, and feminists who
argued girls should be trained to pass the baccalaureate and be able to pursue university studies. In the pages of the *Revue Universitaire*, the moderate feminist Jeanne
Crouzet Benaben, in particular, championed the cause of equal access to degrees.
By 1914 all the Parisian lycées had unofficially added Latin to their curriculum,

and finally in 1924 the Bérard Law extended the public secondary program for girls from five to six years and introduced an optional curriculum that allowed girls to prepare for the baccalaureate.[27] By 1930, the "feminine" curriculum had virtually disappeared. Still separate girls' institutions thrived, as the concept of separate but equal schooling satisfied many French pedagogues and particularly school head-mistresses who defended the "feminine" environment of their schools. Nonetheless, for some, and particularly those women with an awareness of international debates, the issue of coeducation and separate institutions for boys and girls would become an issue.

The "Problem" of Coeducation

Although the revolutionary philosopher Condorcet had advocated coeducation back in 1791, the presence of both sexes in secondary schools was unthinkable throughout the nineteenth century, despite reports from travellers that such arrangements worked to good effect in the United States.[28] When Camille Sée introduced his law, he noted as well that coeducation existed elsewhere, but in France the relationship between the sexes seemed to preclude such a solution. Catholic hostility to coeducation played into French reticence, but more profoundly observers believed that French society was not prepared for such an audacious move. Despite this reticence, however, there is evidence that families were not always as conservative as educational policy. When a girls' *collège* or *lycée* was not in the vicinity, they were prepared to place their adolescent daughters with boys. One even finds examples of mature young women seeking to attend the final years of a *collège* in order to pass the baccalaureate. This was the case for Mlle Marie Suzanne Magin, for example, who petitioned successfully at the age of twenty-nine to attend the classes in philosophy at the *collège* of Ste Ménéhould in Eastern France in order to receive the training necessary to pursue university studies.[29]

During the interwar period, feminists, Catholics, and pedagogues debated the virtues or dangers of coeducation, positioning themselves with respect to the Pius XI's hostility toward "the coeducation of the sexes" expressed in his 1929 encyclical, *Divini illius magistri*. In practice, however, it seems as if coeducation in secondary education progressed mainly for practical reasons, as opportunities for educated women expanded in the service sector.[30] By 1939, educational statistics revealed that, out of a total of 75,000 girls attending secondary schools, 13,000 (17 percent) were in boys' secondary schools.[31] The pragmatic acceptance of coeducation was probably stronger among the lower social groups who traditionally did not have access to secondary education—the latter only became free in 1930 some fifty years after the provision of free schooling in primary schools. When the *lycée* Marcelin Bertholot opened outside of Paris in Saint-Maur in 1938, it was coeducational from the outset, although girl students were a small minority. By 1945, however, some 39 percent of the school's clientele were girls, who came, far more than their male counterparts, from artisanal or employee families more familiar with the primary than the secondary system.[32] In general, it appears that the experience of the Second World War encouraged the development of coeducational schooling.

Future historian Annie Kriegel recalled attending a boys' *lycée* in Cayeux since their family did not return to occupied Paris:

> The classes were mixed of course. It had not even been contemplated that the rigorous segregation that was the rule in all French *lycées* at the time could be maintained in this improvised school. The daring change passed unnoticed and did not give rise to any change in the way we spoke or behaved.[33]

This "daring change" was soon to become the norm, since in the space of thirty years after the end of the war the entire French educational system would become coeducational from kindergarten to the final years of the *lycée*.

Postwar Realignments: Democratization and Coeducation

Despite women receiving the vote in 1944, the immediate postwar period was characterized by widespread social conservatism and a strong domestic message addressed to women. While motherhood and modern kitchens dominated the feminine press, girls pursued secondary studies in increasing numbers, and many used the baccalaureate to enter the university as well. By 1959 young women constituted 56 percent of the student body in literary studies and approximately 35 percent in the other faculties (science, medicine, and law).[34] This move into higher education was a response not only to a hunger for greater and more specialized knowledge among young women, but also to the growth of professional opportunities for women, domestic messages notwithstanding. The more dramatic changes in educational terms occurred at the secondary level where massive restructuring created a system that was no longer divided so strongly along class terms. While democratization was the proclaimed and far more trumpeted goal of these realignments, coeducation accompanied the emergence of a common secondary school for pupils between the ages of twelve and fifteen—the *collège unique*.

Briefly presented, laws and decrees between 1957 and 1976 implemented coeducation at all levels of the educational system for primarily economic and pragmatic reasons. In the context of the postwar baby boom and the need for qualified technicians, secondary schools lost much of their elite character as the required school-leaving age went up to sixteen in 1959 and more and more students progressed from primary school to the *collège*. Coeducation accompanied these dramatic changes in the structure of the French educational system since it was far cheaper to open just one school for both sexes than to continue to provide separate schooling for adolescents. The ruling that first announced this shift to coeducation had no feminist overtones, but argued that the growth of coeducational secondary education was in order "to serve families in their immediate vicinity or in the best possible pedagogical conditions." Financial considerations were, of course, the bottom line. In 1963 the new *collèges d'enseignement secondaire* (lower secondary schools) were coeducational from the outset and the decrees implementing the Haby Law of July 11, 1975 extended coeducation to the *lycées* as well. These educational measures were passed in a society rocked by the cultural revolution associated with the protests of 1968;

undoubtedly the sexual "revolution" of these years and the rebirth of a feminist movement contributed to making coeducation and the assumption that young girls would have the same educational opportunities as their brothers commonplace. By 1971 more girls than boys had the baccalaureate.

The introduction of coeducation attracted little attention among the numerous other changes affecting secondary schools for the masses. These years also witnessed the end of school uniforms, rigid disciplinary measures, and student rankings. While interwar headmistresses had feared for their jobs if coeducation were to prevail, these fears were largely unwarranted. The expansion of the economy meant school teaching became less attractive to men, while for women it represented a career that allowed them to juggle the dual demands of a profession and a family. As a result, the feminization of the secondary school profession continued to progress, reaching a peak in 1971 when approximately 60 percent of all *lycée* professors were women. Still, the number of women school directors declined, particularly among the more prestigious *lycées*.[35] Unquestionably, adolescent girls were the winners in these transformations, more than the socially underprivileged boys who entered the secondary system for the first time in these postwar years. By the early 1990s, sociologists of education were highlighting the differential rates of success between boys and girls at all levels of the educational system.[36] Although gender discriminations remained rife in employment, girls were more successful than boys in obtaining their baccalaureate.

Conclusion

The history of girls' secondary education in France reveals the weight of gendered cultural norms in determining women's access to knowledge and the characteristics of this knowledge. Unquestionably, compared to other countries, French women benefitted from the legacy of Old regime polite society where cultured aristocratic women held sway. As a result, when girls' institutions multiplied in the nineteenth century, this model continued to hold a certain attraction, as reflected in the development of a curriculum that included literature, history, and the sciences alongside religion and the accomplishments. In many ways the enduring competition between religious and lay models of girls' secondary education contributed to efforts to expand the curriculum and especially to provide more rigorous teacher training. The long tradition of single-sex institutions where girls from the middle classes spent a few years in boarding schools acquiring feminine "talents" contributed, however, to a system where girls' education evolved with little relationship to boys' education. Here the existence of feminist voices was decisive in noting that for women to have an impact in civil society they needed an instruction that paralleled, if not equaled, that of men's.

Notes

1. See Mona Ozouf, *Women's Words: Essay on French Singularity* (Chicago: University of Chicago Press, 1997).

2. Françoise Mayeur, *L'enseignement secondaire des jeunes filles sous la IIIe République* (Paris: Presses de la Fondation Nationale des Sciences Politiques, 1977).
3. Françoise Mayeur, *L'éducation des filles en France au XIXe siècle* (Paris: Hachette, 1979; 2nd ed. Perrin 2008).
4. Christina de Bellaigue, *Educating Women. Schooling and Identity in England and France, 1800–1867* (Oxford: Oxford University Press, 2007); Rebecca Rogers, *Les demoiselles de la Légion d'honneur: Les maisons d'éducation de la Légion d'honneur au XIXe siècle* (Paris: Plon, 1992; 2nd ed. Perrin 2006); and Rebecca Rogers, *From the Salon to the Schoolroom. Educating Bourgeois Girls in Nineteenth-century France* (University Park, PA: Pennsylvania State University Press, 2005). See, as well, my historiographic essay: "L'éducation des filles: Un siècle et demi d'historiographie," *Histoire de l'éducation* 115–116 (2007): 37–79.
5. See Rogers, "The Politics of Writing the History of French Girls' Education," *History of Education Researcher* 80 (November 2007): 136–144.
6. Martine Sonnet, *L'éducation des filles au temps des Lumières* (Paris: Cerf, 1987).
7. A few exceptions to this dearth of studies on the twentieth century include: Henri Peretz, "La création de l'enseignement secondaire libre de jeunes filles à Paris (1905–1920)," *Revue d'histoire moderne et contemporaine* 32 (April–June 1985): 237–275; Marlaine Cacouault, *Professeurs… mais femmes. Carrières et vies privées des enseignantes du secondaire au XXe siècle* (Paris: La Découverte, 2007); Loukia Efthymiou, *Identités d'enseignantes—Identités de femmes. Les femmes professeurs dans l'enseignement secondaire public en France (1914–1939)*, thèse d'histoire, Université Paris 7, 2002 and the issue on "Coéducation et mixité," dir. F. Thébaud and M. Zancarini-Fournel, *Clio. Histoire, Femmes et Société* 18 (2003).
8. Carolyn Lougee, "Noblesse, Domesticity, and Social Reform: The Education of Girls by Fénelon and Saint-Cyr," *History of Education Quarterly* 14, 1 (Spring 1974): 87–113.
9. See the table concerning the social origins of the Ursulines of Rouen in Roger Chartier, Marie-Madeleine Compère and Dominique Julia, *L'Éducation en France du XVI au XVIIIe siècle* (Paris: SEDES, 1976), 235.
10. See Sonnet, *L'éducation des filles*, especially, 196–201, 231–261.
11. Jean-Jacques Rousseau, *L'Émile, ou de l'éducation* (Paris: Flammarion, 1966), 475.
12. Mme de Lambert, *Avis d'une mère* (1728), Mme de Miremont, *Traité de l'éducation féminine*, 7 vol. (1779–1789). See Samia Spencer, "Women and Education," in *French Women and the Age of Enlightenment*, ed. S. Spencer (Bloomington: Indiana University Press, 1984), 83–96.
13. Isabelle Brouard-Arends and Marie-Emmanuelle Plagnol-Diéval, eds., *Femmes éducatrices au siècle des Lumières* (Rennes: Presses universitaires de Rennes, 2007).
14. Elke Harten and Hans-Christian Harten, *Femmes, culture et Révolution* (Paris: Éditions des Femmes, 1989), 460.
15. Catherine Montfort and J. Terrie Quintana, "Mme Campan's Institute of Education: A Revolution in the Education of Women," *Australian Journal of French Studies* 33 (January–April 1996): 30–44.
16. *The Journals and Letters of Fanny Burney (Madame d'Arblay)*, vol. 5, *West Humble and Paris, 1801–1803*, ed. Joyce Hemlow et al. (Oxford: Clarendon Press, 1975), 368.
17. For details on the evolution of secondary education for girls in the nineteenth century, see Mayeur, *L'éducation des filles* and Rogers, *From the Salon*.
18. See Rogers, *Les Demoiselles*.
19. Napoleon I, "Notes sur l'établissement d'Ecouen" (1807); the document is reproduced in full in French in Rogers, *Les Demoiselles*, 332–335.
20. See Phil Kilroy, *Sophie Barat: a Life* (Mahwah, NJ: Paulist Press, 2000).
21. See Bellaigue, *Educating Women*, 200–230; Rogers, "French Education for British Girls in the Nineteenth Century," *Women's History Magazine* 42 (October 2002): 21–29.
22. Rebecca Rogers, "Boarding Schools, Women Teachers and Domesticity: Reforming Girls' Education in the First Half of the XIXth Century," *French Historical Studies* 19 (Spring 1995): 153–181.
23. Rogers, *From the Salon*, 164.
24. Félix Dupanloup, *La femme chrétienne et française; dernière réponse à M. Duruy et à ses défenseurs par Mgr l'évêque d'Orléans* (Paris: Charles Duniol, 1868), 108.

25. Jo Burr Margadant, *Madame le Professeur: Women Educators in the Third Republic* (Princeton: Princeton University Press, 1990). See Mayeur, *L'enseignement secondaire des filles*, for details on the genesis of the law, the growth of institutions, and the social characteristics of both students and teachers.

26. Mayeur, *L'enseignement secondaire des filles*, 378.

27. See Karen Offen, "The Second Sex and the *Baccalauréat* in Republican France, 1880–1924," *French Historical Studies* 3 (Fall 1983): 252–288.

28. See Albisetti in this volume, as well as Rebecca Rogers, ed., *La mixité dans l'éducation. Enjeux passés et présents* (Lyon: École Normale Supérieure Éditions, 2004).

29. Archives Nationales, AJ[16] 8679, administrative correspondence between April 15 and April 22, 1929.

30. See Linda Clark, *The Rise of Professional Women in France: Gender and Public Administration since 1830* (Cambridge: Cambridge University Press, 2000).

31. Marilyn Mavrinrac, "Conflicted Progress: Coeducation and Gender Equity in Twentieth-Century French School Reforms," *Harvard Educational Review* 67 (Winter 1997): 772–795.

32. Cécile Hochard, "Une expérience de mixité dans l'enseignement secondaire à la fin des années 1930: le lycée Marcelin Berthelot à Saint-Maur-des-Fossés," *Clio. Histoire, Femmes et Société* 18 (2003): 113–124.

33. Annie Kriegel, *Ce que j'ai cru comprendre* (Paris: Laffont, 1991), 115, cited in Siân Reynolds, *France Between the Wars: Gender and Politics* (London and New York: Routledge, 1996), 50.

34. Claude Lelièvre and Françoise Lelièvre, *Histoire de la scolarisation des filles* (Paris: Nathan, 1991), 125–126.

35. Marlaine Cacouault-Bitard, "La féminisation d'une profession est-elle le signe d'une baisse de prestigé?" *Travail, Genre et Sociétés* 5 (March 2001): 93–115; and *Professeurs... mais femmes*. In 1999, women represented 56.4 percent of the teaching corps in secondary schools and 32.5 percent of directors.

36. Christian Baudelot and Roger Establet, *Allez les filles!* (Paris: Seuil, 1992).

Chapter 3

The Influence of Confession and State: Germany and Austria*

Juliane Jacobi

The German states and parts of the Austrian Empire were known for their efforts to establish a public school system as early as the eighteenth century. It is difficult, however, to produce a consistent narrative about the German schools, since the absence of a substantive central authority in educational matters is one of the peculiarities of the German-speaking states in the nineteenth and twentieth centuries. In the wake of the splitting of the Austrian Empire in 1866 and the foundation of the German Reich in 1870, the historiography has come to reflect Germany's federal character and Austria-Hungary's multinational composition. This chapter will concentrate on the characteristic features of the history of girls' secondary education, examining major differences by focusing on Bavaria, the German-speaking provinces of Austria, and Prussia. Bavaria, a state with an overwhelmingly Catholic population, shares some major traits with Austria, while Prussia represents Protestant Germany.

The German historiography on girls' secondary education began in the early twentieth century and was confined to the territory of Imperial Germany, as it existed after 1870. Jakob Wychgram published a history of girls' secondary education in 1901,[1] which was followed by Gertrud Bäumer's article on Germany and Auguste Fickert's on Austria in the *Handbuch der deutschen Frauenbewegung* (Handbook of the German Women's Movement).[2] While the first half of the twentieth century witnessed no major work on the subject, in 1966 Elisabeth Blochmann, the first German woman to hold a university chair in education, made a new attempt to tell the story of girls' secondary education.[3] Peter Petschauer published a study of the early modern period in 1989.[4] James Albisetti's *Schooling German Girls and Women* offers an in-depth view of secondary schooling for girls during the long nineteenth century.[5] He relies predominantly on Prussian sources, with references to other German states. More recent work pays tribute to the differences among the German states and regions.[6] Gertrud Simon and Margret Friedrich published histories of girls' secondary education in Austria including biographies of outstanding women in the field of girls' secondary education, and reflecting the activities of the Austrian women's movement.[7] A study of middle-class female culture in Austria devotes special attention to nineteenth-century women teachers.[8] Biographies of the women's movement activists Helene Lange and Gertrud Bäumer contextualize their educational efforts within the modernization of the German state and society.[9] In the two-volume history of girls' and women's education coedited by Elke Kleinau and Claudia Opitz, more than sixty authors cover a wide range of issues from the history of ideas to individual institutions and biographies.[10] In its three volumes on the eighteenth and nineteenth centuries, the *Handbuch der deutschen Bildungsgeschichte* (Handbook of German Educational History) features one chapter dealing with girls' secondary education, while the volume covering 1918–1945 integrates the topic into the chapter on schools and their transformation.[11] Recent studies have reconsidered the role of the teaching orders and their contribution to girls' secondary education in Bavaria, Austria, and other German states.[12] The volume *Sozialgeschichte und Statistik des Mädchenschulwesens in den deutschen Staaten 1800–1945* (Social History and Statistics of Girls' Schools in the German States, 1800–1945) represents a major historiographical achievement, since it offers a wide range of information on the development of girls' secondary schools.[13]

The Eighteenth Century: Women's Nature and Place

From the sixteenth century on, girls attended public elementary schools in Protestant regions and schools run by religious women in Catholic regions. Girls' schools had existed to a certain extent in many German towns since the Reformation, but their curriculum did not extend beyond the elementary level. Boarding schools founded by Catholic teaching orders in the seventeenth century served the upper classes and wealthy Catholic families, while well-to-do Protestants preferred to educate their daughters at home, if at all.[14] Beyond the elementary level, instruction focused on domestic subjects such as housekeeping and needlework. Toward the end of the eighteenth century, a number of outstanding new girls' schools were founded, many of them connected with the Enlightenment movement for pedagogical reform.

The improvement of girls' education was part of the lively debates on how to educate children and advance schools, typical of the "pedagogical" century. These debates proceeded from enlightened ideas about human nature, which almost universally posited that women's nature differed from men's. In Germany, as elsewhere in Europe, Jean-Jacques Rousseau's novel *Emile* enjoyed most influence among philosophers of education. The objectives of Emile's education were autonomy and public service, while Sophie's complementary skills were conceived as loving submission and motherliness in service to the family. Among the countless treatises on male and female nature and their implications for the education of middle- and upper-class girls that followed *Emile* in Germany,[15] Joachim Heinrich Campe's *Väterlicher Rat für meine Tochter* (Paternal Advice to My Daughter) demonstrates a reception of Rousseau that entailed even greater restrictions on girls' education. Campe not only emphasized the hierarchy between men and women, but also paid far more attention than Rousseau to preparing girls for their domestic role, banishing intellectual content from female education.

In accordance with Rousseau's initial model of gender relations, women's nature was associated with intuition and sentiment and men's with reason and moral superiority. This "natural" difference between men and women and their place in public or at home was transformed and institutionalized in the nineteenth-century nation-building process. In Germany, as elsewhere in Europe, both men and women of the German middle and upper classes widely accepted women's social role as loving wife, caring mother, and competent homemaker, which left its imprint on girls' schooling.

From the Late Eighteenth Century to 1914

Although school reforms were implemented almost everywhere in the German-speaking countries in the aftermath of the French Revolution, approaches to girls' education differed markedly between Catholic and Protestant countries, both in terms of who was in charge of schools and whether men or women were considered proper teachers for girls. In the German lands, the anticlerical criticism that had emerged all over Catholic Europe during the final decades of the eighteenth century did not survive the post-Napoleonic second decade of the new century. As a consequence, religious girls' secondary schools dominated in states like Bavaria and Austria, while in Protestant Prussia girls' schools were operated by trustees, female schoolteachers, or municipalities.

Catholic Nation-Building

"How can a nun driven to the convent by unfulfilled love, despair or fantasy offer a proper education to a girl whose destiny is to keep her husband and family happy?" wondered Aloys Gscheider in a 1782 pamphlet published in Vienna and entitled *Sind Ordensgeistliche und Nonnen, die in Schulen die Jugend unterrichten, dem Staate wirklich mehr nüzlich als schädlich?* (Are clerics and nuns teaching school more useful than dangerous?). Gscheider joined the warnings of enlightened reformers throughout Europe who questioned the competence of religious orders to teach school. How did religious orders manage to maintain nearly complete control over schools for girls of all social classes in Catholic states in the nineteenth century? Although in Bavaria, as in most Catholic countries, the state took over secondary schools for boys at the end of the eighteenth century, it made no serious effort to do the same for the few girls' secondary schools that had existed since the seventeenth century, and instead reorganized the girls' convent schools. Anticipating high future costs under unfavorable fiscal conditions, the state responded with caution to the demands for girls' public secondary education made by parents, local authorities, and reformers in the new century.

In 1806, the government dissolved eight of fourteen orders altogether and consolidated the remainder into six communities, one for each Bavarian district. As for the capital Munich, the king and his minister Maximilian von Montgélas (1759–1838) erected two secondary schools for girls funded almost entirely from the expropriated holdings of the Institute of Mary or *Englische Fräulein*. The *Königliche Erziehungsanstalt für die Töchter höherer Stände (Max-Josef-Stift)* (Royal Institute for the Daughters from Higher Ranks) was established in 1813. The model for this new school, whose director Thérèse Chardoillet came from Paris, was the French boarding school Écouen.[16] The teachers were hired by the state and accorded the privileged status of civil servants. The curriculum differed from the convent boarding schools only in its stronger emphasis on foreign languages. Until the first headmistress retired in 1830, most subjects were taught in French, the language of the aristocracy. Some subjects, such as languages and religion, were taught by male teachers, underlining the school's exceptionally high status. The sixty girls were chosen from the highest military and civil service families, creating a clear class division between the pupils of this establishment and those of the second, less exclusive, boarding school for girls under royal patronage. In 1817 the *Weibliche Erziehungsanstalt in Nymphenburg für die Töchter mittlerer Stände* (Royal Girls' Institute for Daughters from the Middle Classes) succeeded the school of the *Augustiner Chorfrauen*, who had been teaching in a side wing of Nymphenburg castle since 1731. Following a moderate enlightened curriculum with German as the language of instruction, the new foundation at Nymphenburg educated sixty daughters of urban middle-class families and higher civil and military servants. As an expression of their official status, these schools were open to all daughters of civil servants, whether Catholic or Protestant. The *Stetten Institut* (1805), a similar type of school for Protestant girls in Augsburg, was made possible by a private endowment.

In Bavaria, plans to erect girls' public secondary schools in every district failed and girls' education deteriorated rapidly until the early 1820s. Only in Munich did a strong and demanding city council successfully apply for state funding for a girls' public secondary school (*Höhere Töchterschule*, founded 1822). In the newly acquired Bavarian provinces, which were either Protestant or mixed Protestant and Catholic, Nuremberg was the only city to erect a secondary girls' school of the same type, in 1823.

The revival of Catholic education, which took place all over Europe after 1815, led in Bavaria to the readmission of nuns as teachers in the girls' public elementary and secondary schools. After 1825, Ludwig II reconstituted the estate of the religious orders. In the case of the teaching orders, the local authorities were instructed to supervise the congregations' use of funds for girls' schools. The state controlled the number of novices the orders could admit, strictly regulated the age at which young women entered the order, and sought to ensure that they only joined voluntarily. Most of the orders now accepted girls from the middle- and upper-middle classes and reopened boarding schools. Except for occasional dynastic sponsorship, this educational sector received no public funding, since the schools were regarded as a means of financial support for the religious orders.

The reestablishment of the *Englische Fräulein*, who had taught in Munich since 1625, had a particularly strong impact on the development of girls' secondary

education in Bavaria. They enjoyed an excellent reputation as teachers of middle-
and upper-middle-class girls in the 1820s when five institutes were reopened in
Bavaria. The *Englische Fräulein* reached the peak of recognition in 1835 when
Ludwig I placed the public secondary girls' school in Nymphenburg under their
direction. Unlike orders organized in autonomous branches, which had to keep
their members on site, the *Englische Fräulein* benefited from centralized organization with a chief superior who could handle personnel with great flexibility. They
did not practice enclosure and were able to transfer teachers from their houses in
England, Ireland, and Italy to Germany and vice versa in order to maintain a high
standard of foreign language teaching. In Bavaria there were fourteen institutes with
fifty-eight branches in 1873. The 2,230 members included 685 teachers instruct-
ing 2,800 middle- and upper-class girls in secondary boarding and day schools and
13,790 in elementary schools.[17]

Only one congregation surpassed the activities of the *Englische Fräulein*.
The *Arme Schulschwestern* (School Sisters of Notre-Dame), founded by Theresa
Gerhardinger in 1837, was a community of women who taught girls' elementary
school and became the largest and most influential congregation in Bavaria and
later in other German states and overseas. After 1837, the *Arme Schulschwestern*
began operating many of the Bavarian teacher training institutes, in keeping with
the 1836 teachers' certification regulations.

Schools for the daughters of higher-ranking military officers and civil servants
were founded under royal patronage in many of the German states during the late
eighteenth and early nineteenth centuries, often to train governesses. Some survived the nineteenth century, while others, lacking funding and potential pupils,
soon vanished.[18] The Austrian government had established two institutions of the
same type about three decades earlier, during the time of the enlightened reformer
Joseph II: The *Offiziers-Töchter-Pensionat* (Military Officers' Daughters' Boarding
School) (1775) established to train governesses from the upper classes for the upper
classes and the *Civil-Mädchen-Pensionat* (Civil Servants' Girls' Boarding School)
(1786), which became the nucleus of the women's teacher training institution.

Despite the enlightened reforms of Joseph II in Austria, in 1806 a law concerning public schools returned all authority in elementary school affairs to the church.
Since girls were clearly not expected to attend public school beyond the elementary level, this meant that girls of all social classes were educated predominantly by
religious orders, who operated elementary day schools for lower- and middle-class
girls and boarding schools for the daughters of well-to-do and noble families. The
Ursulines were particularly active in setting up schools, and dominated the field
of training female elementary teachers. Apart from two endowed schools for the
daughters of the military officers and civil servants, some private secular secondary
schools were also established by female teachers or trustees.

Protestant Nation-Building

Like Bavaria and Austria, Prussia put little effort into girls' secondary education
when establishing its public school system. The *Luisenstiftung* (Luise foundation) in

Berlin (1811), fully funded by the king, educated a few governesses aged eighteen to twenty-four in a three-year course. The *Königliche Elisabethschule* (1827) and *Neue Töchter Schule* (1832), later known as *Königliche Augusta Schule* (1863), offered an eight-year program for middle-class girls. Both schools had male principals and an overwhelmingly male faculty. Contrary to initial expectations, they received no major financial support from the State or the royal family and depended almost exclusively on fees from parents.[19]

In the provinces, dynastic support was absent and girls' education beyond the elementary level relied on parental self-help. Municipal support, private enterprise, or trustees provided some sort of secondary education for middle-class girls. In the industrial town of Bielefeld, parents took the initiative and established the *Höhere Töchterschule* in 1828. The municipality took responsibility in the 1840s, since the town's leading liberal democrats were among its supporters. In 1858, an association of conservative upper-middle-class Protestants founded the Lutheran private *Töchterschule*. Women ran the school until 1913, whereas the municipal *Höhere Töchterschule* always had a male principal and a predominantly male faculty.[20]

In 1861, private girls' schools in Prussia outnumbered public schools 345 to 229,[21] but they were far smaller in size. The average student-teacher ratio was fewer than twelve per teacher in private schools, as compared to forty in public schools.[22]Private schools varied in size and profile from municipal schools, most were female-run, and women frequently outnumbered men on the faculty. Some of these initially private foundations were eventually transferred to municipal responsibility, and as a result nearly all these schools directed by women were soon taken over by male principals. Parental choices apparently depended on class, income, and religious affiliation. Only the well-to-do could afford the conviction that girls should be taught by women in small family-like groups. Bielefeld's small but visible Jewish community had to send their daughters to the municipal *Höhere Töchterschule*, while the leading Christian businessmen and industrialists sent their daughters to the private *Töchterschule*. In larger cities, the Jewish community established its own girls' secondary schools, but throughout the century many Jewish girls attended public schools.[23]

From 1827 to 1864, Prussia had between 250 and 350 girls' public schools, which meant that every large and even many smaller towns maintained a girls' secondary school. Like the above-mentioned Berlin institutions, these schools had male principals and, with the exception of the first three elementary grades (*Unterstufe*), an exclusively male faculty. Only in provinces with a predominantly or at least large Catholic population (i.e., Rhineland, Westphalia, Silesia, and Posen) were girls' public secondary schools rare. Where they existed, women teachers outnumbered men.[24]

During the first decades of the nineteenth century, Protestant women teachers were compelled to acquire their professional knowledge on their own. Following the advent of the certification examination for male elementary teachers in 1826, the State introduced in 1837 an examination for female teachers in the province of Brandenburg, which was gradually extended to the other provinces and accompanied by the foundation of teacher training institutes for women—Protestant and Catholic, private or attached to public schools.

During the first half of the nineteenth century, private girls' schools placed a strong emphasis on instruction in fine needlework, music, and other so-called aesthetic subjects, which were deemed indispensable female accomplishments. Public secondary schools provided some training in needlework, but generally speaking their curriculum more closely approximated that of boys' secondary schools.

Curriculum guidelines existed on the provincial or local level. In 1869, the prescriptions for girls' schools stated that they should prepare pupils for women's destiny and duties, that is, family life. Subjects to be taught included religion, reading, German language and writing, natural history, geography, history, singing, drawing, arithmetic, French, and English, the latter only in the highest grade. Such schools should admit children from the elementary level on and generally keep them longer than children usually spent in school. They were to provide elementary, middle, and upper level education. Given these rather vague guidelines, what happened between 1868 and 1908, when what Albisetti has called the "decisive reforms in female education" took place, putting girls' secondary education on an equal footing with that of boys?

Accelerated Growth and Delayed Reforms: Prussia and Austria from 1860 to 1912

After 1860 a number of concurrent developments promoted girls' secondary education. During the 1860s and 1870s, many private and public secondary schools for girls responded to the growing shortage of male teachers at both elementary and secondary level by adding a teachers' seminary.[25] In 1874, when Prussia banished nuns from teaching as a result of the anti-Catholic policies of the *Kulturkampf* (struggle for culture), the State had to found new teachers' seminaries for Catholic laywomen to replace them, while other states such as Bavaria and Württemberg established their first nondenominational seminaries for girls' secondary school teachers.

Important support for upgrading girls' schools came from the male directors of girls' secondary schools and from the women's movement. The principals of girls' schools—many of them university-educated and, since 1872, organized professionally—promoted the upgrading of girls' public secondary schools and hiring more academically trained faculty in order to raise their own professional status. They explicitly supported the idea of the truly feminine character of girls' secondary education and opposed adopting the boys' *Gymnasium* curriculum and the access to careers it offered.

The women's movement began campaigning for middle-class women's right to work outside the home during the 1860s. Teachers were strongly represented, and demands for enhanced educational and career opportunities led in 1890 to the founding of the *Allgemeine Deutsche Lehrerinnenverein* (General German Women Teachers' Association) headed by Helene Lange (1848–1930). Lange became the leading figure not only in the female teachers' movement, but also in the campaign to improve girls' secondary schools and gain access to the universities for women.

Unlike the male teachers' association, Lange believed that women should teach girls' secondary school. She argued that it was imperative to improve academic training for female teachers in order to upgrade girls' education. Given the State's reluctance to open up academic careers for girls, in 1893 Lange introduced a privately financed four-year course to prepare graduates of the *Höhere Töchterschule* for the *Abitur* examination. In Prussia, most initiatives followed Lange's proposal, while some women's associations in other German states favored a six-year academic course beginning earlier, at age thirteen or fourteen, after seven years of schooling.[26] The State reacted with new regulations in 1894, but provided no public options to prepare for the *Abitur*. Female teachers, however, now had access to an academic certificate that allowed them to teach the upper grades in girls' secondary schools.

As in many other European countries at the turn of the century, growing pressure from the women's movement and social reformers underlined the need to afford at least some girls the opportunity to prepare for the *Abitur* and enter university. In 1908, the ministry of education enacted new regulations, establishing a ten-year girls' school (*Lyzeum*) that offered no academic or career preparation. Graduates could then continue with a two-year program in home economics or a three-year teacher-training course, which eventually (1912) opened the possibility of enrolment in university studies for secondary school teachers. Three secondary paths leading to the *Abitur*, known as *Studienanstalten* (institutions for study), were introduced. Girls could take a six-year course, branching off after seven years of *Lyzeum*, which mirrored either the *Gymnasium* (Latin and Greek) or *Realgymnasium* (Latin and more modern languages and science). A five-year track similar to the *Oberrealschule* (no ancient languages) branched off after eight years. Finally, girls' secondary schools were incorporated into the public school system.

In Austria, the *Reichsvolksschulgesetz* (Imperial Elementary School Law) of 1869 abolished church control over elementary schools and opened *Bürgerschulen*, which went beyond the elementary level, to girls. For the first time, the Austrian State recognized female teachers who were not nuns in the public schools. As a consequence, by 1896 more than 40 percent of Viennese elementary school teachers were women.[27]

Like its German counterpart, the Austrian women's movement began campaigning for the advancement of girls' schools in the 1860s. In Vienna, the *Frauenerwerbsverein* (Women's Employment Association) opened a four-year secondary course for girls in 1872. It was soon extended to six years, which resembled the first privately funded *Mädchenlyzeum* in Graz. In 1885, the latter became the first municipally run *Lyzeum* in Austria. Similar foundations followed in other cities. While the State granted girls access to the *Matura* examination as early as 1872, university philosophical faculties did not permit regular enrolment until 1897/98. Not until 1900 could women acquire academic degrees as secondary school teachers. Academic girls' schools were founded far earlier in some of the non-German-speaking parts of the Austrian Empire.

Even after the final reform of 1912, when the *Mädchenlyzeum* was entitled to expand its six-year lower-level program with a four-year course preparing for the *Matura*, the number of state or municipally sponsored girls' secondary schools did not grow significantly, although most existing schools were expanded to become *Reformrealgymnasium* (similar to Prussia, but with Latin started later). This meant that girls' secondary schools could become *Mittelschulen*, the Austrian term for secondary schools preparing for higher learning rather than the term "middle schools" in some German states. With the exception of a few municipal schools, most girls' secondary schools were run by religious orders, trustees, and entrepreneurs.

Mission and Colonialism

The religious revival in the aftermath of the Napoleonic wars led to the foundation of a number of Protestant congregations dedicated to serving the poor, teaching at the preschool and elementary level, and nursing. As part of the European powers' imperial expansion, these so-called deaconesses established girls' boarding schools abroad that offered education beyond the elementary level. Three branches of the Kaiserswerth congregation were established in the Ottoman Empire. They founded boarding schools for the daughters of middle-class families in the multicultural Levantine ports of Smyrna (1860) and Beirut (1851), and in Jerusalem (1851). In comparison to the French Catholic congregations with whom they competed both nationally and confessionally, the deaconesses' teaching missions abroad did not flourish in the long run.[28]

German Catholic congregations pursued the interests of Imperial Germany to a lesser degree. Estranged from the hegemonic Protestant leadership of the newly created *Reich* after 1870, they were preoccupied with resisting the pressures placed upon them during the *Kulturkampf*. It was only in the twentieth century, when German nuns became part of the international missionary activities of the Catholic Church, that they founded new mother houses in South America, South Africa, and the Far East.[29]

The Twentieth Century

From the First World War to 1945: Ambivalent Modernization in Girls' Secondary Schools

After the integration of girls' secondary schools into the public school system in Prussia and elsewhere in Germany in the aftermath of the First World War, secondary schools were still divided into *Gymnasien* and *Oberschulen*, which prepared for the *Abitur* (and thus gave access to higher education), and *Mittelschulen*. *Mittelschulen* offered a middle-level examination that opened a number of vocational training possibilities. In an ongoing process of change from 1910 to 1938, the German secondary school system acquired its peculiar tripartite structure. After a compulsory four-year public elementary school (*Grundschule*), regulated in 1919, pupils were channeled into three distinct types of secondary education, of which the Volksschule did not grant any career opportunity, and only the *Gymnasium* and *Oberschule* would provide access to higher learning. In West Germany until the 1970s, this kept advanced secondary education leading to the *Abitur* highly selective and discouraged broader access by girls.

The reorganization of elementary school teachers' training between 1919 and 1925 led to major changes in girls' secondary education. In 1919, Prussia and most other German states began closing teachers' seminaries for men and women and establishing academic teacher training colleges requiring the *Abitur*. This meant that the *Lehrerinnenseminare* attached to girls' secondary schools were converted into preparatory courses for the *Abitur*. The educational system of the first German Republic continued, in general, to distinguish between the curricula of boys' and girls' secondary schools.

Nevertheless, the regulations of some German states such as Baden and Württemberg had always allowed girls to attend boys' secondary schools when girls' schools were unavailable. During the Weimar Republic, this became a much more common practice in all German states except Prussia. In a number of states, between one-third and one-half of girls attending secondary courses leading to the *Abitur* were enrolled in boys' schools.[30]

The six-year course of the *Lyzeum*, the counterpart to the boys' *Realschule*, maintained an explicitly gendered curriculum. It prepared girls to enter a wide range of female occupations requiring its lower-level diploma. These vocational options included nursing and health care, kindergarten teaching, clerical training (*Höhere Handelsschule*), teaching needlework and physical education (*technische Lehrerin*), and home economics.

In Austria, girls were admitted to boys' *Gymnasien* beginning in 1919/1920. Girls' classes were to be established if demand permitted. A special *Frauenoberschule* offering a gender-specific curriculum was introduced in 1921 on the initiative of female secondary school teachers opposed to coeducation, many of them religious women. The *Frauenoberschule* prepared for matriculation in only a limited range of university courses.

With rising unemployment and a revival of anti-intellectualism (including complaints of "too many students from the wrong background unsuited to academic studies") and antifeminism in educated circles,[31] the final years of the Weimar Republic witnessed a public debate on the utility of girls' secondary schooling and women's university studies, in the wake of which female university enrolment dropped. In many German states, after 1930 government policy discriminated against women and only male secondary teachers were hired.[32] In this spirit, the newly established Nazi *Reich* Ministry of Education enacted a "Law against the Overcrowding of Secondary Schools and Universities" in 1933. The main targets of this law were Jewish students, who were successfully banned from enrolling in most universities over the next two years. The law also established quotas for women.

After 1935, however, nearly full employment and the demand for more university graduates led female enrolment to rise again.[33] At the same time, however, the ideological discrimination against women who wished to prepare for the *Abitur* and university studies did not vanish, but rather became an intrinsic part of Nazi educational propaganda. Four principles of Nazi policy affected girls' secondary education: the consolidation of the different types of secondary schools in the German *Reich*, the suppression of private secondary schools run by religious orders, the proliferation of secondary schools promoting a gender-specific curriculum and career choices, and the public suppression of women's demands for academic training. The major changes in the schools took place in 1936/1937. The *Deutsche Oberschule für Jungen* (German Upper School for Boys), initially created to replace the preparatory courses for male elementary school teachers, became the standard secondary school for boys, while the *Gymnasium* with a core curriculum of Greek and Latin was considered a "special form." The *Gymnasium* course was no longer offered for girls, and the *Oberschule für Mädchen* was the only secondary school left to prepare them for the *Abitur*. The upper grades of the *Oberschule* were divided into a language course and a home economics course informally dubbed the *Puddingabitur*. These schools were part of an ideological system that denied women's claims to equal educational opportunities and channeled many girls' into gender-specific careers for the rest of their lives.

After the corporatist *Ständestaat* regime took over in Austria in 1934, coeducation was abolished there and it became more difficult for girls to enter the *Gymnasium*. In 1938 the Nazis imposed the German school system, offering girls a language course or a home economics course. Religious orders were expelled from the country and their schools either closed down or taken over. Reopened after the war, their dominance in girls' secondary education went unchallenged during the first decades of the new Austrian Republic.

Reform and Expansion from 1945 to 1990

Postwar Germany was divided into two states in 1948/1949. The FRG (Federal Republic of Germany) became a member of the Western alliance, while the GDR (German Democratic Republic) was integrated into the Communist bloc. In most West German states this meant reconstructing the prewar tripartite school system, while in the GDR a comprehensive school system was gradually established. There, a centralized system of school governance was set up along with a ten-year course for all children, followed by a selective two-year program to prepare students for the *Abitur*. The reform was completed in 1959 and codified in 1965. No further substantial reforms took place until the breakdown of the socialist German state in 1990. Consolidating the various school systems in the federal *Länder* (states) of the FRG has been a constant task for educational leaders up to the present day. When many West European countries began to open their secondary schools for the majority of pupils, West German public opinion not only became worried about the selectiveness of its educational system, but also found itself in competition with East Germany.

In the 1950s and 1960s, East German policies aimed to give equal opportunities to girls and made a strong effort not only to increase the number of girls in secondary schools, but also to change the "bourgeois" ideal of gendered occupations and channel girls into technical professions such as engineering and the sciences. Coeducation was introduced immediately after the war at all levels, while most secondary schools in the FRG were single-sex.

Women's share of university enrolment, which reached 47.5 percent in 1977, appeared to confirm the success of the GDR's equal opportunity policy. At the same time, fears arose that girls would outnumber boys, and the authorities were instructed to maintain "equal gender relations." A closer look reveals that the authorities themselves limited or even counteracted the success of these policies. Although the proportion of women in technical occupations in the GDR rose from 5 percent in 1962 to 25 percent in 1989, gender differences did not vanish under state socialism. In 1975, women outnumbered men in teacher training courses (by 73 to 27) and in the arts and humanities (by 72 to 28). Even in medicine (by 66 to 34), women outnumbered men at an early stage, since little financial incentive was granted to medical doctors by the state-run health care system.

These ratios reveal that throughout the existence of the GDR, women's career choices were far more limited to gender-specific occupations than were men's.[34] In the FRG, girls were the largest segment among the three underprivileged groups sociologist Ralf Dahrendorf identified in his widely read 1965 pamphlet *Bildung ist Bürgerrecht* (Education as a Civil Right). While some 50 percent of all primary school pupils were girls, they made up only 41 percent of those accepted to secondary schools preparing for the *Abitur*. Only 36 percent of girls completed secondary school, and when it came to university enrolment, only 26 percent of students were women. Among university graduates, only 17 percent were young women. The higher the level of education, the lower was the proportion of girls participating.[35]

Although both post-1945 German states witnessed a gross gender imbalance resulting from two world wars, social and educational policies differed remarkably. In West Germany, nineteenth-century notions of gender persisted, and with them the assumption that the majority of girls were unsuited to serious scholarly work or science. The continuous stream of well-educated refugees from the GDR up to the construction of the Berlin Wall in 1961 limited demand for highly educated women, and ideological hostility to GDR educational and social policy reinforced traditional patterns. West German social policy implicitly rejected women's employment, and tax regulations and family allowances bolstered a gender-specific division of labor based on the nineteenth-century middle-class ideal. Gender disparities in the educational system and the disadvantages girls faced were not issues of public debate until the 1960s, and only female educators in the girls' secondary schools spoke out against discrimination and for a gender-specific idea of girls' secondary education in line with the interwar women's movement.[36]

While Dahrendorf and other social scientists such as Helge Pross[37] were still musing over how to counteract the obvious gender bias in educational opportunities, a silent change was already underway among parents and pupils. Girls' enrolment in secondary schools had increased gradually since the early 1950s, and the number of girls attending *Realschule* (the equivalent of the *Mittelschule*; formerly *Lyzeum*) outnumbered boys from 1950 onward. In 1968 the total number of girls in secondary schools exceeded boys, and by 1972 girls even gained the majority in the *Gymnasium*.[38]

The late 1960s witnessed a major change in the formerly elitist *Gymnasium*. The number of female graduates, which had risen from 32.2 percent in 1950 to 37.3 percent in 1967, further increased to 45.9 percent in 1975. By the end of the decade, girls made up 50 percent of *Gymnasium* graduates. This amazing shift did not mean that differences and discrepancies between boys and girls in secondary schools vanished. Of the three streams, girls chose the modern language option far more often than boys did. Only a very small proportion of girls pursued the classical language stream, and only 14 percent of girls (compared to 37 percent of boys) attended the science stream.

At the same time, the concept of the *Gymnasium* changed rapidly in the wake of the reform debates of the 1960s and 1970s. The debate over comprehensive secondary schooling influenced the development of the *Gymnasium*, since a growing number of pupils were attending and graduating from that type of school. The silent incursion of girls into the *Gymnasium* was accompanied by a rapid shift in the secondary schools from sex segregation to coeducation, which took hold in nearly all schools in the early 1970s, except those run by Catholic orders. In Austria as well, secondary coeducation became mandatory in 1975.

Yet the *Gymnasium* continued to be male-dominated, since most of the principals as well as the majority of the faculty were men. The 1980s witnessed a feminist debate on coeducation and male bias in the school administration and faculty. The male-oriented *Gymnasium* curriculum also became controversial. Women did come to comprise 50 percent of the faculty, but since more work part-time, the *Gymnasium* remains a male-dominated institution. Disadvantages deriving from gender stereotyping, which shapes teachers' perceptions of boys and girls, also became part of the pedagogical debate. While much research has been done on coeducation's impact on students of both sexes, there has, interestingly enough, been very little study of the relationship between the teacher's gender and girls' future choices and achievements.

Conclusion

In the German-speaking countries, girls' secondary education was long marked by the religious divide that has shaped German history since the Reformation. In the early nineteenth century, girls' secondary education in the Catholic regions or states was almost exclusively in the hands of religious orders, which were supported (or expropriated) by the state. While policies instituted after the foundation of the German Empire in 1870 repressed the activities of teaching orders there, nuns continued to dominate girls' secondary education in Austria until after the Second World War. In Protestant countries, the State took almost no interest in girls' secondary education until the last decade of the nineteenth century. Here, post-elementary education for middle- and upper-class girls was provided either by private institutions operated by female lay teachers or municipal public secondary schools (*höhere Töchterschulen*). The latter primarily employed male lay teachers, while female personnel taught only at the elementary level.

After 1870, the women's movement fought to improve girls' secondary education. At the same time in Prussia and the other predominantly Protestant states of Imperial Germany, the male headmasters of girls' public secondary schools sought official recognition of their institutions within the state's public secondary school system. The decisive reforms of girls' secondary education took place in the first decade of the twentieth century, putting girls' secondary education on a nearly equal footing with boys'. In Austria, public secondary girls' schools developed at

Table 3.1   Chronology of girls' education in Austria, Bavaria, and Prussia

Austria	Bavaria	Prussia/German Empire/GDR/FRG
1775: Offiziers-Töchter-Pensionat		
1786: Civil-Mädchen-Pensionat	1805: von Stetten Institute Augsburg	
	1806: Religious orders dissolved	
	1813: Max-Josef-Stift	1811: Luisenstiftung in Berlin
	1817: Weibliche Erziehungsanstalt in Nymphenburg	
	1822: Munich Höhere Töchterschule	
	1825: Reconstitution of religious orders	1827: Elisabethschule in Berlin
		1832: Neue Töchterschule
	1836: Teaching certificate for women	
	1837: Arme Schulschwester founded	1837: Teaching certificate for women
1869: Secular female teacher training courses		1861: 229 public secondary schools for girls
1873: First Mädchenlyzeum in Graz		
1877: Frauenerwerbsvereinschule		1874: Catholic schools closed, religious orders expelled
		1890: Allgemeine Deutsche Lehrerinnenverein founded
		1893: Courses preparing for Abitur in Berlin;Mädchengymnasium in Karlsruhe (Baden)
1897/98: Admission to philosophical faculties		1894: New regulation for girls' secondary schools and female secondary school teachers
1900: Admission to medical School	1903: Admission to universities	After 1902: Universities gradually accepted female students
1912: Mädchenlyzeum with options for Matura	1911: Girls' secondary schools put on a par with boys' schools	1908: Girls' secondary schools put on a par with boys' schools
1919: Admission to law school		
1938: Private (i.e., religious) schools closed		1933: "Law Against the Overcrowding of Secondary Schools and Universities"
		1946: GDR coeducation in all public schools
1975: Coeducation in public schools mandatory		After 1970: FRG coeducation in all public schools

a slower pace. It was not until 1872 that the first Austrian public school for girls (*Mädchenlyzeum*) was founded, and courses preparing for the *Matura* exam required for university matriculation gained little support from the state even during the first Austrian Republic (1918–1934).

The upgrading of teacher training in Germany after the FirstWorld War promoted the replacement of seminaries for female teachers by eight-year courses preparing for the *Abitur*. This reform altered the structure of girls' secondary schools and a growing number of girls passed the *Abitur* exam. At the same time, the enrolment of girls in schools offering six-year courses (known variously as *Mittelschule, Realschule,* or *Lyzeum*) expanded substantially.

Nazi ideology and practices were inconsistent if not contradictory, with the curriculum of girls' secondary schools becoming more gender-specific while the number of women entering university studies rose substantially.

After the Second World War, the East German state, while granting girls equal opportunities in coeducational secondary schools, failed to abolish gender disparities in career opportunities. After ten years of comprehensive school, the majority of girls were channeled into gender-specific occupations in the service and care sectors, preschool and elementary teaching. In higher education, men outnumbered women in the sciences and engineering.

In West Germany, gender-specific ideas on education affected girls' secondary education for a far longer period. Only in the 1970s, when coeducation was introduced in all public secondary schools, did girls' enrolment in schools preparing for the *Abitur* match that of boys (see table 3.1).

Notes

*I would like to thank Pamela Selwyn for helping with the English version.

1. Jakob Wychgram, "Geschichte der Höheren Mädchenerziehung," in *Geschichte der Erziehung,* ed. Karl Adolf Schmid, vol. V, part 2 (Stuttgart: J.G. Cotta, 1901).

2. Gertrud Bäumer, "Geschichte und Stand der Frauenbildung in Deutschland," and Auguste Fickert, "Der Stand der Frauenbildung in Österreich," in *Handbuch der deutschen Frauenbewegung,* vol. III: *Der Stand der Frauenbildung in den Kulturländern,* ed. Bäumer and Helene Lange (Berlin: W. Moeser, 1902), 1–128, 160–191.

3. Elisabeth Blochmann, *Das "Frauenzimmer" und die "Gelehrsamkeit"* (Heidelberg: Quelle und Meyer, 1966). This study was conceived as the first volume of a larger study and covers only the eighteenth and early nineteenth centuries.

4. Peter Petschauer, *The Education of Women in Eighteenth-Century Germany* (Lewiston, NY: Mellen, 1989).

5. James C. Albisetti, *Schooling German Girls and Women. Secondary and Higher Education in the Nineteenth Century* (Princeton: Princeton University Press, 1988).

6. Studies of other German states and Austria as well as local histories of larger cities offer additional insights into these differences as well as a broader perspective on the impact of religion. On Austria, see Gertrud Simon, *Hintertreppen zum Elfenbeinturm: Höhere Mädchenbildung in Österreich. Anfänge und Entwicklungen* (Vienna: Wiener Frauenverlag, 1993); and Margret Friedrich, *"Ein Paradies ist uns verschlossen...": Zur Geschichte der schulischen Mädchenerziehung in Österreich im "langen" 19. Jahrhundert* (Vienna: Böhlau, 1999); on Baden, Rupert Kubon, *Weiterführende Mädchenschulen im 19. Jahrhundert am Beispiel des Großherzogtums Baden* (Pfaffenweiler: Centaurus, 1991); and Rosemarie Godel-Gaßner, *Die Geschichte der mittleren*

Mädchenbildung in Baden und Württemberg von 1871 bis 1933: Ein Beitrag zur allgemeinen Entwicklungsgeschichte der baden-württembergischen Realschule (Frankfurt a.M.: Lang, 2004); on Bremen, Martina Käthner, *Der weite Weg zum Mädchenabitur: Strukturwandel der höheren Mädchenschulen in Bremen (1854–1916)* (Frankfurt a.M.: Campus, 1994); on Bavaria, Christel Knauer, *Frauen unter dem Einfluss von Kirche und Staat: Höhere Mädchenschulen und bayerische Bildungspolitik in der ersten Hälfte des 19. Jahrhunderts* (Munich: Kommissionsverlag Uni-Druck, 1995); on Württemberg, Karin De la Roi-Frey, "Schulidee: Weiblichkeit. Höhere Mädchenschulen im Königreich Württemberg 1806–1918," Diss. University of Tübingen 2004 (URL: http://w210.ub.uni-tuebingen.de/dbt/volltexte/2004/1353/); on Hamburg, Elke Kleinau, *Bildung und Geschlecht: Eine Sozialgeschichte des höheren Mädchenschulwesens in Deutschland vom Vormärz bis zum Dritten Reich* (Weinheim: Beltz, 1997); and on Lübeck, Sylvia Zander, *Zum Nähen wenig Lust, sonst ein gutes Kind...: Mädchenerziehung und Frauenbildung in Lübeck* (Lübeck: Archiv der Hansastadt Lübeck, 1996).

7. Simon, *Hintertreppen*; Friedrich, *Paradies*.

8. Gunda Barth-Scalmani, "Geschlecht: weiblich, Stand: ledig, Beruf: Lehrerin. Grundzüge der Professionalisierung im Primarschulbereich in Österreich bis zum Ersten Weltkrieg," in *Bürgerliche Frauenkultur im 19. Jahrhundert*, ed. Brigitte Mazohl-Wallnig (Vienna: Böhlau, 1995), 343–400.

9. Angelika Schaser, *Helene Lange und Gertrud Bäumer. Eine politische Lebensgemeinschaft* (Cologne, Weimar and Vienna: Böhlau, 2000); Juliane Jacobi, "Helene Lange," in *Klassiker der Pädagogik*, vol.1, ed. Heinz-Elmar Tenorth (Munich: Beck, 2003), 199–215.

10. Elke Kleinau and Claudia Opitz, eds., *Geschichte der Mädchen- und Frauenbildung*, 2 vols (Frankfurt a.M.: Campus, 1996).

11. *Handbuch der deutschen Bildungsgeschichte*, vols 2–4, ed. Christa Berg, August Buck, Christoph Führ, Carl-Ludwig Furck, Notker Hammerstein, Ulrich Herrmann, Georg Jäger, Karl-Ernst Jeismann, Dieter Langewiesche, Peter Lundgreen, Detlef K. Müller, Karlwilhelm Stratmann, Heinz-Elmar Tenorth, und Rudolf Vierhaus (Munich: C.H. Beck, 1987–1991).

12. See Ilse Brehmer and Gertrud Simon, eds., *Geschichte der Frauenbildung und Mädchenerziehung in Österreich: Ein Überblick* (Graz: Leykam, 1997); Christel Knauer, *Frauen unter dem Einfluss*; Juliane Jacobi, "'Entzauberung der Welt' oder 'Rettung der Welt': Mädchen- und Frauenbildung im 19. Jahrhundert," *Zeitschrift für Erziehungswissenschaft* 9 (2006): 171–186.

13. Bernd Zymek and Gabriele Neghabian, *Sozialgeschichte und Statistik des Mädchenschulwesens in den deutschen Staaten 1800–1945*, Datenhandbuch zur Deutschen Bildungsgeschichte, vol. 2: Höhere and mittlere Schulen, part 3, ed. Detlef K. Müller (Göttingen: Vandenhoeck & Ruprecht, 2005).

14. For Protestant Germany, see Juliane Jacobi, "Zwischen 'nöthigen Wissenschaften' und 'Gottesfurcht': Schulische Mädchenbildung von der Reformation bis zum 18. Jahrhundert"; for Catholic Germany, see Andreas Rutz, "Das Primat der Religion. Zur Entwicklung separater Mädchenschulen in den katholischen Territorien des Reiches im 17. Jahrhundert," both in *Säkularisierung vor der Aufklärung? Bildung, Kirche und Religion 1500–1750*, ed. Hans-Ulrich Musolff, Juliane Jacobi, and Jean-Luc Le Cam (Cologne: Böhlau, 2008), 253–274; 275–288.

15. See, among other works, Claudia Honegger, *Die Ordnung der Geschlechter. Die Wissenschaft vom Menschen und das Weib* (Frankfurt and New York: Campus, 1991); and Christine Mayer, "Geschlechteranthropologie und die Genese der modernen Pädagogik im 18. und 19. Jahrhundert," and Elke Kleinau, "'Der Mann der Weibes Herr auf den Stufen der Kultur.' Bemerkungen eines aufgeklärten Zeitgenossen über Geschlechterverhältnisse im Kulturvergleich," both in *Bildungsgeschichten: Gender, Religion und Pädagogik in der Moderne*, ed. Meike Baader, Elke Kleinau, and Helga Kelle (Cologne, Weimar and Vienna: Böhlau, 2006), 119–140, 141–158.

16. See chapter 2 on France in this volume

17. Knauer, *Frauen unter dem Einfluß*, 321.

18. Albisetti, *Schooling German Girls*, 27–30.

19. Ibid.

20. Ilse Brehmer and Juliane Jacobi, eds., *Frauenalltag in Bielefeld* (Bielefeld: AJZ, 1986).

21. Zymek and Neghabian, *Sozialgeschichte und Statistik*, 40.

22. The statistics are somewhat questionable. Zymek and Neghabian, *Sozialgeschichte und Statistik*, 39.

23. Andreas Hoffmann, *Schule und Akkulturation: geschlechtsdifferente Erziehung von Knaben und Mädchen der Hamburger jüdisch-liberalen Oberschicht 1848 – 1942* (Münster: Waxmann, 2001).

24. Zymek and Neghabian, *Sozialgeschichte und Statistik*, 41.

25. Albisetti, *Schooling German Girls*, 67; Zymek and Neghabian, *Sozialgeschichte und Statistik*, 30–33.

26. The first *Mädchengymnasium* was founded in Karlsruhe, Baden, in 1893.

27. James C. Albisetti, "Female Education in German Speaking Austria, Germany and Switzerland," in *Austrian Women in the Nineteenth and Twentieth Century*, ed. David Good, Margret Grandner, and Mary Jo Maynes (Providence and Oxford: Berghahn 1996), 39–57.

28. Malte Fuhrmann, *Der Traum vom deutschen Orient: Zwei deutsche Kolonien im Osmanischen Reich 1851–1918* (Frankfurt a.M.: Campus, 2006), 118–126, 145–151, 206–208, 231–235; Julia Hauser, "'Herzensbildung im christlichen Sinn': Erziehungsarbeit Kaiserswerther Diakonissen im Osmanischen Reich (1851/60–1918)," unpublished manuscript, Wittenberg, 2008.

29. Of the graduates of Württemberg's nondenominational teacher's seminary class of 1899, four young women joined a religious order and three of them became mothers general of congregations in Brazil, South Africa, and Japan, respectively. See *Freundschaft in sieben Jahrzehnten: Rundbriefe deutscher Lehrerinnen 1899–1968*, ed. H. Jansen (Frankfurt a.M.: Fischer, 1991), 337–341.

30. Zymek and Neghabian, *Sozialgeschichte und Statistik*, 77 and 79, Table 4.3.2, cols. 4 and 5.

31. Konrad Jarausch, *Deutsche Studenten 1800–1970* (Frankfurt a.M.: Suhrkamp), 129–139.

32. Claudia Huerkamp, *Bildungsbürgerinnen. Frauen an den Universitäten und in akademischen Berufen* (Göttingen: Vandenhoeck & Ruprecht, 1996), 116–117.

33. Jarausch, *Deutsche Studenten*, 202–204.

34. Heidemarie Kühn, "Madchenbildung im Schulsystem der DDR," in Kleinau and Opitz, *Geschichte*, 2: 434–445, 442.

35. Ralf Dahrendorf, *Bildung ist Bürgerrecht. Plädoyer für eine aktive Bildungspolitik* (Hamburg: Zeitverlag, 1992), 72.

36. Torsten Gass-Bolm, *Das Gymnasium 1945–1980: Bildungsreform und gesellschaftlicher Wandel in Westdeutschland* (Göttingen: Wallstein, 2005), 148.

37. Helge Pross, *Über die Bildungschancen der Mädchen in der Bundesrepublik Deutschland* (Frankfurt a. M.: Suhrkamp, 1969).

38. Peter Lundgreen, *Sozialgeschichte der deutschen Schule im Überblick*, Part II: *1918–1980* (Göttingen: Vandenhoeck & Ruprecht, 1981), 115–116.

Chapter 4

Chequered Routes to Secondary Education: Italy*

Simonetta Soldani

Secondary schooling has always been a crucial component in building individual and collective identities based on shared knowledge and values; nonetheless, it has received little attention in Italian historiography. Even studies of the *ginnasi* and *licei*, central to the education of the nation's ruling classes, are few;[1] the situation is still worse for technical, vocational, and teacher training schools.[2] The prevailing approach has favored pedagogical and philosophical debates over models of secondary schooling or investigation of institutional vicissitudes. Scant attention has been devoted to its actual protagonists—the teachers, the head teachers, the pupils, who for us today are unfamiliar figures.[3] This is even more the case for girls, who only began to appear in postprimary schooling in Italy in the second half of the nineteenth century, because of a widespread lack of interest and die-hard reticence about the suitability of giving lower-middle-class women (and even those of bourgeois and noble families) knowledge that went beyond the basics, sometimes bolstered by lessons in etiquette, drawing, music, and dance.

In the late sixteenth century, as a result of the Catholic Reform, and later with occasional sovereigns' interventions, *collegi* (private boarding schools) and *convitti*[4] began to appear for boys, while absolutely nothing was done for girls until much later. Throughout the eighteenth century, the education of girls from well-to-do families meant several years of stay in a convent, where they seldom learned more than reading, writing, and arithmetic, and often even these to quite a sketchy standard. "Enlightened" fathers concerned with their daughters' education were still a rarity. The few women given access to universities to study—like Maria Pellegrina Amoretti, who graduated "*in utroque iure,*" that is, in both canon and civil law—or to "hold lessons"—like the famous mathematician Maria Gaetana Agnesi—were rare oddities.[5] The handful of women called upon to spice up poetic contests with their "improvised" verse or to perform in private academies and public theatres

were also exceptional, and their lifestyles were too far removed from the norm to
represent a socially acceptable model. Even Madame de Staël confirmed this opinion, condemning the cultured and free poetess Corinne—almost a projection of
Italy—to lose first the man she loved, and then, prematurely, her life.[6]

To overcome this "exceptionalist paradigm," schooling for girls had to become a
value not just for single individuals, but for the female sex as a whole, for the families
in which women were encapsulated (in thought and reality), and for the national
community to which they "belonged." This occurred as part of an epoch-making,
transnational movement of ideas, eager to redesign real and symbolic, individual
and collective identities. On this basis, the female inhabitants of the "country called
Italy" were urged to cast aside the superstitions and prejudices, fatuity and immorality that prevented them from becoming the "citizens' wives and mothers." In fact,
a reform of private and public customs, an indispensable premise to any project for
the "resurrection (*Risorgimento*) of the nation," was unthinkable without the active
participation of women.[7]

Women's Education for the "Risorgimento" of the Nation

It is hazardous to speak of true secondary education for girls in Italy before 1860.
Not only were there very different levels of female culture in the various regions,
but even the few French-style *maisons d'éducation* for noble and rich girls founded
in the Napoleonic era—in Milan (1808), Verona (1812), Lucca (1807), and Naples
(1809)—financed by the State and run by lay personnel specifically dedicated to
the girls' education, did not truly qualify as secondary, given the modest amount
of schooling they provided.[8] The spectrum of knowledge imparted by the Istituto
della SS. Annunziata di Firenze (1825) was not much broader, except for the prestige
derived from its teaching the "national" language, which originated in Tuscany.[9]

As for the over 500 conservatoires, orphanages, *educandati*,[10] and *ritiri* (retreats)[11]
scattered about the peninsula, hosting female boarders and day pupils and run by nuns
and oblates, none provided an education that went beyond reading, writing, and arithmetic. Even the various female religious congregations who tried to open girls' educational institutes faced distrust of schooling for women from society and the ecclesiastic
authorities, who still considered it not only a pointless but even a harmful exercise.[12]

Nevertheless, starting in the 1830s, in northern Italy and the largest towns a
significant decline in well-off families sending their daughters away to a convent for
their education coincided with an increasing willingness to spend money to endow
them with some culture. At the highest levels this occurred in the home; but more
girls learned the rudiments of literature and history, French, geography, geometry,
and introductory natural sciences in small private and quasi-domestic schools. This
type of education developed, for example, at the *Istituto Wulliet* in Livorno, the
school of Carolina Nencioni and Enrichetta Lambert in Florence, and the *Istituto
Desnisard* and the Elliot sisters' school in Turin. Such establishments were only open

to girls under fourteen, however, since respectable girls could not continue to attend a day school beyond that age.[13]

The 1848 revolution complicated the picture further. Between 1846 and 1849 the hope to "provide women with a sound education" came to be increasingly stained with liberal and civil overtones, as many patriots maintained that the "Italian" woman could not remain solely "the guardian and maker of domestic order and well-being."[14] It was necessary to bind girls to exacting and strict studies, so that the mind could educate the will, "discipline" the passions, and render women of use to the common fatherland, as Caterina Franceschi Ferrucci wrote on the eve of revolution in her treatise *Della educazione morale della donna italiana*. This highly successful work, reprinted several times, aspired toward studies for middle-class girls that could include Latin, Law or Philosophy, or even the "History of World Civilisation."[15]

It was improbable that this level of schooling could be attained. The failure of the *Collegio Nazionale delle Peschiere* for Italian girls in Genoa confirmed the challenges. Established in 1850 by a handful of patriots and exiles and placed in the hands of Franceschi Ferrucci, this school was rejected by the very fathers for whose daughters it had been designed. However, in the 1850s there was a sharp increase in the demand for postprimary school education for middle-class girls, despite explicit hostility both on the part of the Church and the sovereigns. Only Piedmont undertook to create a network of modest provincial "methods schools" for boys and girls that prepared for the "royal licence," required of primary school teachers since 1846.[16]

Initial Explorations (1860–1875)

With their minds set on developing "national" middle classes to entrust with state administration, the Italian ruling classes concentrated their attention and limited funds available on boys' secondary schooling, indispensable in making the "different peoples" of Italy, "who neither knew nor understood each other, a single people."[17] That choice also aimed to shape the backbone of the nation through schooling's "assimilative virtues." As for girls, excluded from knowledge and power, considered "only ideally" as citizens,[18] prevented from exercising professions and taking up employment, and trapped—in the middle classes—in the home, the State had little concern with their education. The sole exception was again the State's commitment to create *scuole normali* for boys and girls to train the primary school teachers needed to implement the compulsory schooling envisioned by Piedmont's Casati Law of 1859, which was applied to other regions after unification.[19]

At that time, however, it was difficult to consider normal schools, conceived of as vocational courses, as "secondary," although they were the only lay, state, and free schools available for families who either could not or would not spend on private schools or various "educational Institutes for girls." These institutes, almost all religious, seldom strayed from the holy triad of prayer, catechism, and women's domestic tasks; many also refused to recognize the new State and its emblems.[20] There was no alternative for girls unless they or their families had the courage to "knock on the doors of the boys' schools" to demonstrate by their example the "normality" of "co-attendance by scholars of different genders," as hoped for in 1866 by Cristina Princess of Belgiojoso, mindful perhaps of her youthful Saint-Simonian experiences.[21]

The ruling classes' failure to build channels for educating young middle-class women was particularly shortsighted. Indeed, they could have made these women active supporters of the new order of things, still threatened by the all-powerful Catholic Church's refusal to recognize the Italian national State. Yet, many politicians shared the widespread prejudices against education for women, although open-minded men in the government did appoint a few "female patriots"—journalists and writers—to teach in prestigious girls' schools (such as Luisa Amalia Paladini in Florence, Anna Maria Mozzoni in Milan, and Erminia Fuà Fusinato in Rome). However, even this simple action was a first step toward real and symbolic enhancement of the "nation's women," wholly unthinkable just a few years earlier. Even a decree from 1867 aiming to secularize the girls' conservatoires inherited from past regimes, particularly numerous in Tuscany and Sicily, aroused fierce opposition from the local nobility and the public they influenced. At the same time an attempt begun by the municipalities to promote upper girls' schools—three-year day schools called *Scuole femminili superiori*—only took off in a few towns in central-northern Italy. Moreover, enrolment in these schools peaked in 1874 at a few hundred girls. This failed attempt to create secondary schools aimed at providing a "culture" specifically for girls would not be the last.

However, at that time something as unexpected as it was important for the future of girls' schooling was happening: growing numbers of girls enrolled at the *normali*, which, thanks to the broader syllabi launched by the Ministry for Education in 1867, began to attract "not only female pupils whose intent was to become primary school teachers, but young ladies from well-off families wishing to procure some culture at not too great a cost."[22] The ratio of *scuole normali* for girls and boys rose from 21–20 in 1863 to 36–23 in 1873. A larger gap developed with the *magistrali*, shorter duration teacher training schools established by local authorities: forty-three for girls versus thirteen for boys. In addition, the girls' normal schools achieved far better results: inspectors described swarms of young ladies who with "the zeal and fervour of all novices…pay careful and assiduous attention to the lessons in profound silence, punctually and accurately complete their homework, and read with curiosity and the desire to learn."[23]

Only in the south did "the ancient and traditional revulsion" against female education persist even "among the higher citizenry." Even the generous subsidies set aside for pupils who were "mainly orphans" or barely capable of reading were not enough to fill the girls' *scuole normali*.[24]

Testing the Ground, Tracing Pathways (1875–1895)

The expansion of normal schools had several indirect consequences: (1) the establishment of one-year, then two-year, "preparatory courses" (the so-called *scuole complementari*) to fill the "void in schooling" between the end of primary school and the *normali* (which girls started at fifteen); (2) admission of girls graduating from the *normali* to the university courses required for teaching in postprimary schools (not full matriculation); and (3) the establishment in Rome and Florence in 1882 of the *Istituti Femminili Superiori di Magistero* (upper girls' teacher training institutes). In addition, the *normali* were becoming the institutions "that laid down the rules and the norms for all the others," as "the teaching at many *convitti* was starting to be modelled on them" and even "the few female schools bearing the name of *superiori* do not differ a great deal from them."[25] In 1880, a "special envoy" from the French Ministry for Education admired the ability of these schools to benefit the mélange of "young women from different social conditions" who, since they were subject "to the same studies, the same discipline, and the same intellectual and moral influences," developed "a sentiment of collective dignity and self-respect" that he thought few countries could match.[26]

In all likelihood, the real situation was less rosy than these statements indicate. Yet, during the 1880s, the *scuola normale* became more and more "feminine": enrolments grew from around 9,000 girls in 1885 to no less than 22,000 ten years later, aided by the many institutions established by provinces, municipalities, and foundations. Contributing as well was the failure of attempts made between 1879 and 1885 to set up in Italy something similar to the French *lycées de jeunes filles* of 1880,[27] confirmation of the Italian ruling class's ongoing reluctance to invest in schools specifically for girls.

The few officials interested in preventing the education of young Italian women from being monopolized by the Church were wont to underline the political value of the issue. The majority, however, were still convinced that girls should be educated in protected places, under the watchful eyes of nuns who would teach them enough for their "destiny as wives and mothers" and for a modest social life. Indeed, by 1884, there were 1,584 institutes, with almost 50,000 pupils between ages six and seventeen, that had been opened throughout Italy, run by female religious congregations willing to offer less convent-like conditions and teaching under the guidance of a "sacred host" of nuns. These institutes gradually became more attentive to postprimary courses and met with great favor from the old and new middle classes.[28]

Just when deadlock seemed inevitable, news began to circulate of a steady trickle of girls entering classical and technical schools designed for boys. As the law did not explicitly exclude girls from accessing any type of school (perhaps because the lawmakers did not even give it a second thought), headmasters had difficulty closing their schools' doors to the "daring ones" who presented themselves with the backing of fathers who wanted to give them a solid academic schooling at low cost and to entrust their education to schools run by the State for which they had fought and/ or worked (see table 4.1).

Given that in 1875 a royal decree had allowed the entry of women to all the university faculties in the kingdom, which required a *liceo* or *istituto tecnico* diploma, it was inevitable that things would pan out like this. The numbers aiming that high in the beginning were so tiny that the young ladies enrolled in the *licei* and *istituti tecnici* were almost invisible; however, in the course of the 1880s the hope expressed by the Princess of Belgiojoso seemed about to come true. Although protests by some narrow-minded bishops and many conformists against the "undue mixing of the sexes" slowed down the process, by the early 1890s girls in the lower level of the *ginnasi* and *scuole tecniche* already numbered several hundred.[29] Not everyone approved, but the voices defending the right of girls to climb the ladder of knowledge together

Table 4.1   Italian school types, 1859–1923

Track	Name	Years	Type	Comments
Secondary	Ginnasio	3+2	Humanities-based (Latin, Greek, Philosophy)	Two cycles with exit point after 3 years
	Liceo	3		Diploma qualifies for all university faculties
	Scuola tecnica	3		Diploma qualifies for Istituto tecnico
	Istituto tecnico	3, later 4		5 tracks focussing on agriculture, industry, accounting, trade, sciences. The last named qualified for university science faculties
Post Primary	Scuola Normale	3	Teachers' schools	After 1896, it had to be preceded by 3 years of technical schools or scuole complementari (only for girls)
	Scuola superiore femminile	3		
	Scuola complementare	2		Fills gap from end of primary school to normali
Post Primary	Vocational Schools	3+3	Vocational	Various tracks, dependent on Ministry of Economics; in some cases leading to Superior Schools for Agriculture, Engineering, Trade

with their brothers "in order consciously to offer to the country the modest but intelligent contribution of their work" and "to comprehend the laws of their State and the ideals of their Fatherland" were growing louder. They believed that a girl must learn "fully to appreciate the works of human genius" and to feel the need to do so, because "she who has seen the light does not choose to walk in the darkness."[30]

Respectable Young Ladies and Courageous Minorities (1896–1905)

Italy's tormented *fin de siècle* was marred by profound political and institutional malaise, but it also witnessed an unprecedented acceleration in economic and social transformations. Female secondary education experienced, on one hand, an expansion of private religious institutes and, on the other, a rapid increase in enrolments at the public schools "for boys." Only the *normali*—as of 1896 no longer free and requiring postprimary school studies for admission—experienced a decade-long standstill.[31] Conditions at the turn of the century were outlined in a report drawn up by the Ministry for Education for the International Exhibition in Paris in 1900 to illustrate the progress made in "education of the people and women." It recognized that "in Italy the *collegio* and *educatorio* are still the educational institutes most accepted by families as well as the most widespread."[32] The religious educational institutes numbered over 1,000, with 38,000 boarders and 46,500 day pupils, even though the great majority only followed elementary courses.[33] At the same time, the report proudly underlined "with what ease and in how short a time the custom of mixed secondary and university teaching had spread in Italy," and noted this occurred "not only in the Italian regions where there is no prejudice and women are freer to behave as they choose, but also in the provinces of Naples and Sicily, where a few years ago women did not leave the house except to go to church, and even then never totally alone."[34]

Between 1900 and 1905, girls who were enrolled in *scuole tecniche* increased from 3,577 to 9,009, at *ginnasi* from 1,030 to 2,286, at *licei* from 233 to 452, and at *istituti tecnici* from 84 to 513.[35] For the very great majority the experience of going to schools designed for boys was truly significant: they were not just subject to unisex timetables and syllabuses, but also studied in the same classrooms with boys, although at desks placed slightly apart.

State schools for girls only, or with all-girl classes, were a small minority,[36] confirmation of the increasing open-mindedness of families toward coeducation. Many families believed that it would ensure a higher cultural level for their daughters, whereas separate schools and classes would risk "driving them back into an environment of narrow-mindedness and prejudice," as denounced by a combative female teacher from the girls' gymnasium in Milan.[37] Besides, the great majority of (male) teachers also regarded the expansion of mixed classes as a bulwark against the threatening invasion of women secondary teachers, who in 1900 already numbered 870, often using as a springboard for beginning the career subjects and schools of lesser prestige, where salaries were lower and the knowledge required was less formalized.

However, coeducation, "never established by principle," and indeed "introduced as a stopgap and tolerated as transitory,"[38] had no explicit legal legitimization. The public mostly used it for convenience rather than from conviction, as confirmed in the same years by the debate and proposals to reform secondary schools, in which the questions regarding girls' schooling were seldom raised. Indeed, while debate raged elsewhere in Europe, Italian silence played into the hands of those who wanted educational tracks separated by class, gender, and objectives, almost as if modernizing was a byword for forming divisions, specializations, and hierarchies.[39] In an Italy buffeted by the winds of a fragile modernization, the counterattack of "aristocratic" and "masculine" classicism as the main feature of the "thinking nation" produced growing intolerance of young women in search of culture and useful diplomas and stoked new appeals to set up "special schools" just for them.

Like a River in Full Flow (1906–1918)

Never, perhaps, had school-related issues received such intense attention in Italy as in the decade before the First World War. Although the main focus of this attention continued to be boys' education, nonetheless the State's role in secondary education for girls surfaced often at conferences for primary and middle school teachers and head teachers and in major specialized and nonspecialized periodicals (*Nuova Antologia, Rivista Pedagogica, Nuovi Doveri*). There was a particularly intense debate on the *scuola normale*, considered incapable of carrying out the "twofold (cultural and vocational) function" that had been confirmed in regulations of 1896. Educationalists and politicians were invited to learn from other countries, France and Prussia in particular; and the authorities were exhorted to set up state "schools of culture" specifically designed for young ladies from good families, a request often backed by the need to counter the quasi-monopoly of Catholic initiatives in that field.[40] But others thought it pressing that "co-education in the technical, classical, and normal schools be legally recognized,"[41] considering the "good fruits" that the girls had produced in the "technical and classical schools not designed especially for them."[42] The female editor of one of the most popular magazines "for young ladies" of the time asserted that the girls attending gymnasia showed "a more pleasant freedom of movement and thought," their conduct confirming that "the *camaraderie* with their male classmates accustomed them to approaching men without falling into sentimental convulsions at first sight," while in the "girls' private boarding schools" the air was filled with "mysticism, sentimentality [and] selfishness."[43]

While these discussions went on, however, enrolment of girls rose sharply in all types of schools. Including the state and "*pareggiate*" institutions, their numbers at *scuole tecniche* rose from 21,901 to 38,855 from 1911 to 1916, reaching 34.1 percent of total enrolments. At the *ginnasi* the female presence climbed from 5,353 to 11,567, or 23.6 percent. Even the numbers enrolled at the *istituti tecnici* and *licei* were improving: 2,567, or 9.3 percent at the former by 1916 and 1,668, or 13.2 percent at the latter. In the meantime, the *normali* had resumed growing rapidly, from just over 7,000 girls in 1905 to 33,280 in 1916, while the *complementari* reached 30,504.[44]

With over 100,000 girls attending state secondary day schools each year, the private and foundation boarding schools and *convitti* lost relevance, making somewhat less pressing in Italy the great cleft, especially for girls, between the values and lifestyles proposed by the state/lay and private/clerical schools that tore apart French public opinion at the time.[45] An epoch-making change in mentalities and life plans took place as more families in the urban middle classes sought an education for their daughters that both instilled culture "for its own sake" and finished with a diploma that could qualify them for the few paid occupations considered suitable for girls belonging to families free from the vortex of poverty.

These shifting sociocultural strategies could only hinder enrolment in vocational education. This affected institutes dependent on the Ministry of Economics, mostly standard boys' schools, but also those specifically dedicated to preparing for positions "reserved to women in the domestic economy, in cottage industries and special public and private companies: without neglecting... those activities and practices of health education and family economics... indispensable for all mothers."[46] However much stimulus came from the contemporary French example and the woman-in-the-home model, in Italy few families were willing to send their daughters to school to learn how to govern a house, prepare lunch, bring up a child, and receive guests.

The State did need to try to stem the swelling demand for girls' schooling from a civil society that was much richer in dreams of climbing the social ladder than in the material resources and opportunities actually to do so. Already in the troubled prewar period, however, what dominated public discourse on schooling was deep intolerance toward classes and pupils that did not fit with tradition: sons of the lower middle classes, first of all, but also girls, intruders into a culture that was not made for them. Promoters of the idealist wave continually stressed that women should remain "sufficiently close to the deepest sources that the life of the species feeds on." Therefore, an "essentially intellectual education" for women was not only "physiologically and morally inopportune," but also "socially dangerous."[47] Redefining schools for the "aristocracy of the mind and spirit" meant distancing women from them as much as possible. As Giovanni Gentile insisted in spring 1918, women "have not, and never will have, the originality of thought stemming from the soul, nor the steely spiritual vigor which are the superior intellectual and moral forces of humanity, and on which the school for forming the country's superior spirit must rest."[48]

Difficult Years (1919–1927)

The First World War not only exhausted Italy but also greatly disrupted its school system. Irregular attendance, undeserved passes, and easy diplomas, for men as well as for women, occurred in all the warring countries; so did a sharp rise in enrolments at schools preparing for employment in the tertiary sector. But no other "great power" had to face such extremes as Italy, where a very high rate of illiteracy (in 1921, affecting one out of four males and almost one out of three females over the age of six) was countered by incredibly high enrolments of both sexes at secondary schools. In 1920, girls numbered 17,000 of 57,000 students at *ginnasi*, 49,000 of 133,500 at *scuole tecniche*, 3,000 of 6,500 at *licei*, and 7,500 of 35,000 at *istituti tecnici*.[49]

Contributing to this situation was the open nature of the Italian school system, which largely lacked "dead-end" schools, plus its geographical diffusion, necessary to counter the hegemonic objectives of the Catholic schools. In the heated aftermath of the war, disputes focused on these two eternal sticking points, again with infrequent references to the expanding presence of young women. The "assault of female teachers" on secondary schools—their numbers grew from 2,000 to 8,000 during the First World War—drew more attention, especially after approval in 1919 of a law on legal capacity, which allowed women "to exercise all professions and to cover all public and state roles with the same rights as men."[50] After their initial isolation, many women teachers were becoming integral parts of, or even leading, professional and feminist associations, social committees, political movements, and networks of women organized against the sexist policies not only of the government but of many male colleagues, whatever their political beliefs and ideals.

In comparison, condemnation of the legions of young girls every day entering schools not designed for them was relatively rare, even in the southern towns
least willing to admit them. But the extent of the displeasure at their presence
was soon revealed by the guidelines of the "Gentile Reform" (1923), which, while
not excluding women, explicitly declared its desire to reduce their numbers and
importance.

Launched as a Defence of the Realm Act, the reform took effect before parliament briefly discussed it. Subject to criticism for its elitist nature, its creation of
obstacles for educating the lower classes and those living furthest away from the big
cities, and the room that it opened for confessional schools, it was most radical with
regard to the types of schooling that had provided a gateway to secondary education for both girls and the less well-off or less educated classes. The Gentile Reform
transformed the *scuole complementari* into dead-end courses and created a postprimary track with Latin that led to new normal schools known as *istituti magistrali*,
whose numbers were reduced from 153 to 87. It also abolished the existing variety
of *scuole tecniche* in favor of a common preparation (with Latin) for *istituti tecnici*
and transformed their most general course into a *liceo scientifico*. It not only drastically increased fees, including higher tuition for girls than boys, but also placed a
countrywide ceiling on enrolments in the various types of school, with the exception only of the *complementari*. Women were barred from running middle schools as
well as from teaching many disciplines considered unsuitable for them for "cultural"
or civil reasons.[51]

What Mussolini called "the most fascist of reforms," and Gentile liked to present as a "restoration of the Casati law" modernized with ideas and proposals that
had "almost all been debated and proclaimed in Italy before the war,"[52] aimed to
reduce drastically the flow of girls toward the schools they had proven to prefer,
and to palm them off toward the newly established "*liceo femminile*," which gave no
access to university and considered culture more as ornament than substance (see
table 4.2).

But things did not go quite as planned, even though girls' enrolment in secondary schools did plummet from 100,000 in 1915 to just over 60,000 in 1925.[53] The
reduction occurred mainly in less prestigious schools, such as the *complementari*
and the lower level of the technical institute. The teacher training institutes also fell
sharply: in 1925 there were 12,000 girls in the lower courses and 8,000 in the upper,
less than a quarter of the enrolments at the *normali* five years earlier.[54] However,
that girls still supplied 88 percent of prospective teachers offended the promoters of
the reform who wanted to train more male teachers, due to the political importance
attached to primary schools.[55] Moreover, while the much desired "*liceo femminile*"
immediately turned out to be a flop, with the laughable enrolment figure of 122
in 1923 gradually eroding year after year, the *ginnasi* and *licei*, which should have
been at the heart of the "drastic clearing out," saw a much smaller drop in female
enrolments than expected.[56] This trend confirmed the existence of a hard core of
families devoted to cultural—not vocational—education for their daughters. What
is more, on average the girls consistently earned better grades than their male classmates, even now that they were no longer just the best few, but a significant part of
the whole.

Table 4.2   Italian school types, 1923–1938

Track	Name	Years	Type	Latin	Comments
Secondary	Ginnasio + Liceo	5+3	Humanities-based	√ Lower and higher level of school	
	Liceo scientifico	4+4	Modern studies	√ Lower level of school	
	Istituto tecnico	4+4			1st 4 years in common with Liceo scientifico
	Istituto magistrale	4+3	Teachers' schools	√ Lower and higher level	
	Liceo femminile	4+3		√ Lower level	1st 4 years in common with Istituto magistrale; abolished 1928
Postprimary	Scuola complementare	3			Terminal course under MPI; abolished 1929
Vocational	Avviamento al lavoro	3+2	Courses of commerce, manufacturing and craft industries, agricultural activities		Various tracks under MPI and, until 1929–1932, Ministry of Economics
	Avviamento professionale e istituti professionali	3+3 or 5			Upper level vocational courses

Sprinting Ahead (1928–1938)

Early in 1928, the "touch-up policy" begun after Gentile's resignation from the Ministry of Education in July 1924 became a true "demolition" of his reform.[57] The first, thin wall to be knocked down was that of the *liceo femminile*, rejected by even the middle and upper middle classes of society. The second wall to fall was much sturdier: suppressing the controversial *complementari* and putting all the vocational schools with some general education courses under the Ministry for Education. Finally, the tough restrictions on the number of enrolments in teacher training institutes also collapsed, together with the much-publicized separation between boys' and girls' classes.

Between the 1920s and 1930s many aspects of the Fascist regime changed. New needs and new ways of life prompted by the economic crisis or disseminated by films and magazines open to influence from the rest of Europe and America promoted a more fluid educational structure. For example, the urban lower classes welcomed for their daughters the three-year "business training courses" (*scuole di avviamento commerciale*), similar to the beloved *scuole tecniche* of the past, which saw enrolment jump from just over 20,000 girls in 1931 to 47,000 in 1939.[58] Yet the "female vocational schools" that taught cooking and sewing, interior decorating and health, housekeeping and bringing up children continued to stagnate. Though the newspapers liked to repeat that "in the Fascist climate" women should attend schools designed especially to educate them for the "place that nature and civil laws destined for them,"[59] the regime's propaganda could not overcome social resistance. In an Italy shaken by the economic crisis and falling salaries, the majority of the urban working classes preferred that their daughters acquired marketable skills. For middle-class girls, the teacher training institutes (now with Latin) also continued to be a fundamental pole of attraction, as became clear when the Gentile reform's restrictions were abolished and enrolments could climb freely again.[60]

In any case, on the eve of the Second World War the cultured woman of the future was no longer depicted only as the shy aspiring teacher. She was also the dynamic and uninhibited *liceo* girl publicized in the quality magazines, the protagonist in gymnastics competitions and sporting feats, proud of her talents and independence of judgment. Distrust toward a solid cultural training for girls had not vanished, but the climate had changed, as shown by the extraordinary leap from 3,700 girls enrolled at the *liceo classico* in 1931 to over 12,000 in 1939. In the end this trend even overshadowed the large growth in *ginnasio* pupils,[61] who increasingly used that school less to obtain knowledge for its own sake than to gain access to the *liceo*. On this track they measured up to the authors of "masterpieces in art and literature, law-makers and philosophers, scientists, heads of state and captains," that is, to male excellence in diverse fields, as a state education official trying to justify restrictions on female secondary school teachers had written in 1926.[62]

The syllabi certainly were shaped for men, which hampered these robust female pioneers in changing persistent stereotypes of female nature and destiny. In any case, the few schoolgirl memories at our disposal reveal not disillusionment with the lack of "highbrow" models of their own sex but pride in gaining access to the works and lives of those male giants and achieving better results than the boys in schools, which required equal performances and offered equal opportunities.[63] These memoirs also stress that the rare and formal relations between girls and boys of the early 1920s were rapidly shifting toward more relaxed, friendly exchanges.[64]

By 1939, when new regulations substantially modified the Gentile reform,[65] discrimination in access to secondary education was much more based on class than on gender.[66] As always in Italy, one's birthplace also mattered: over the years the gap between town and country, and between North and South, had widened, especially for girls. Those girls who continued postprimary school generally came from higher social levels than did boys, both the over 130,000 enrolled at the secondary schools with Latin and the 110,000 at the various technical, vocational, and training schools. These figures make crystal clear the failure of the clearly misogynous laws of the previous twenty years (see table 4.3).

Table 4.3   Italian school types, 1939–1980s

Track	Name	Years	Type	Latin	Comments
Lower secondary schools	Scuola media	3	Common lower secondary	√	
Upper secondary schools	Ginnasio + Liceo	2+3	Humanities-based	√	
	Liceo scientifico	5	Scientific studies	√	
	Istituto tecnico	5	Accounting and surveying/ engineering tracks		
	Istituto magistrale	4	Teachers' school	√	
Vocational	Avviamento al lavoro	3+2			Mostly under MPI but some under public or private local bodies
	Avviamento professionale e istituti professionali	3+2 or 5			Dependent on MPI

Note: In 1963 the *scuole medie* and lower-level vocational schools were merged into a unified middle school without Latin.

Toward an Oasis of Equal Opportunities?

The advent of a "democratic republic founded on work," as article 1 of the Constitution issued in 1948 reads, did not lead to any significant change in the school system. The strong class connotations continued and, if anything, gender polarization strengthened, with increasingly high numbers of boys at vocational and technical schools and girls more at ease in humanities-based ones and teacher training institutes, where the end of Fascism led the number of boys to plummet from 24,500 to 6,000 over seven years.[67]

Persistent gender stereotypes continued to influence the choice of schooling for Italian families who could afford for their children to pursue postprimary schooling. In 1948, girls comprised 54 percent of students in the three-year course with Latin but barely a third in the vocational training schools. It was good to be prepared to take on "paid work" (not manual) if necessary, but less so to pursue a specific professional qualification, as was required of boys.

The 1950s marked the consolidation of these trends. In 1963 the three-year courses with Latin and vocational training courses were merged into the so-called unified middle school (*scuola media unica*), which finally raised compulsory schooling to eight years (the only important legislative reform of public education in republican Italy). At that time, 600,000 girls attended various postprimary schools, almost 43 percent of the pupils. This amounted to half the female age group between eleven and fourteen, suggesting that even the working classes were prolonging their daughters' schooling, at least in the towns. Yet in a celebrated Catholic school near Florence, pupils reported in 1967 that none of their female classmates had continued after the three years of *scuola media* because "in our village the parents think that a woman can live with a chicken brain."[68] Female enrolments were still struggling to get off the ground in the South, where a massive wave of migration to the cities of northern Italy removed large segments of its younger population. Enterprising and seeking change, such southern women shifted elsewhere the drive for emancipation through schooling that they too desired, for their daughters if not for themselves.

It is worth noting that schooling for the masses became a reality without giving rise to "protected" schools either for the children of middle and upper classes or for girls. Private schools declined precipitously from 21 to 10 percent between 1951 and 1961, then to a meager 5.6 percent ten years later, including closure of hundreds of girls' boarding schools. Almost all survivors were day schools modelled on state schools, despite their tendency to offer courses with a weaker knowledge content. State schools thus absorbed the huge increase of enrolments in secondary education, from 840,000 in 1961 to over 2,500,000 at the beginning of the 1980s, with girls approaching half of all enrolments.[69] Italy's lag behind much of Western Europe had been reduced more quickly than had been thought possible. In the process, women teachers even overtook men in their upper secondary school fortress by the mid-1970s, generating many still unresolved problems.

Equal yet distinct, the young women who crowded the classrooms in all parts of Italy continued to prefer schools offering disciplines perceived as more friendly toward women, such as the teacher training institutes and language-based *licei*, where in the mid-1980s girls accounted for, respectively, 94 and 88 percent of enrolment.[70] But they also numbered two out of three at art institutes and prestigious humanities-based *ginnasi-licei*, nearly half in both the vocational institutes and the *licei scientifici*, and close to 40 percent even in the *istituti tecnici*.

Since then, not much has changed. Italian girls and women have managed to work around social and cultural crises, impossible experiments, and nominal reforms; and they have clearly exceeded the famous fifty/fifty ratio, especially in the more developed regions of the North, where boys stay at school for less time, as they are more concerned with finding well-paid employment or perhaps working in the small family businesses. However, despite higher qualifications, women still lag in employment, pay, and promotions. One gets the impression that the female students of our times are defending a fortress in a phase of abandonment. Is this to be the fate of schools in the much-acclaimed "knowledge society"?

Notes

*Translated by Karen Whittle.

1. See Adolfo Scotto di Luzio, *Il liceo classico* (Bologna: il Mulino, 1999). The *ginnasio* and *liceo*'s syllabuses provided a sound grounding in Latin, Greek and Philosophy: see table 4.1 in this volume.

2. Some important contributions have been published in the *Fonti per la storia della scuola* series by the Archivio Centrale dello Stato.

3. See the "Introduzione" to Simonetta Soldani and Gabriele Turi, eds., *Fare gli italiani: Scuola e cultura nell'Italia contemporanea* (Bologna: il Mulino, 1993); also see journals such as the *Annali di storia dell'educazione e delle istituzioni scolastiche* and volumes edited by La Scuola di Brescia.

4. A *convitto* was a hostel that provided children with primary and secondary education in internal or (more often) external schools.

5. Ilaria Porciani, "Il Plutarco femminile," in *L'educazione delle donne: Scuole e modelli di vita femminile nell'Italia dell'Ottocento*, ed. Simonetta Soldani (Milan: Franco Angeli, 1989), 297–317.

6. Madame de Staël, *Corinne; or Italy* (Oxford: Oxford World's Classics, 1998 [1st ed. 1807]).

7. See Silvana Patriarca, "Indolence and Regeneration: Tropes and Tensions of Risorgimento Patriotism," *The American Historical Review* 110, 2 (April 2005): 380–408; and, from a more specific viewpoint, Simonetta Soldani, "Il Risorgimento delle donne," in *Il Risorgimento* (*Storia d'Italia, Annali* 22), ed. Alberto Banti and Paul A. Ginsborg (Turin: Einaudi, 2007), 183–224.

8. Silvia Franchini, "Gli educandati nell'Italia postunitaria," in Soldani, *L'educazione femminile*, esp. 57 61; Franchini emphasizes their cloister-like discipline and religious customs.

9. Silvia Franchini, *Élites ed educazione femminile nell'Italia dell'Ottocento: L'Istituto della SS. Annunziata di Firenze* (Florence: Olschki, 1993).

10. An *educandato* was a place providing girls with the skills needed for a proper behavior in society and at home.

11. These *ritiri* were convent-like institutions for ladies and girls, who lived there as recluses for years.

12. Cf. Giancarlo Rocca, "Conservatorio ed educandato nell'Ottocento italiano," *Annali di storia dell'educazione e delle istituzioni scolastiche* 2 (1995): 59–101.

13. For the experiences quoted, see Ida Baccini, *La mia vita* (Milan: Unicopli, 2004); Bruno Cicognani, *L'età favolosa* (Milan: Garzanti, 1940); Grazia Pierantoni Mancini, *Impressioni e Ricordi (1856–1864)* (Milan: Cogliati, 1908). For an account of the situation in Florence, defined as "disheartening" compared to Turin and Milan, Silvia Franchini, "Scuola, conservatorio, educandato e tradizioni familiari: l'istruzione femminile a Firenze verso la metà dell'Ottocento," *Annali di storia dell'educazione e delle istituzioni scolastiche* 5 (1998): 165–182.

14. Raffaello Lambruschini, *Sull'utilità della cooperazione delle donne bennate al buon andamento delle scuole infantili per il popolo* (Milan: Stella, 1834), 12.

15. Caterina Franceschi Ferrucci, *Della educazione morale della donna italiana: libri tre* (Turin: Pomba, 1847), 141, 127.

16. Giorgio Chiosso, "Le scuole per i maestri in Piemonte (1840–1850)," *Quaderni del Centro studi "Carlo Trabucco"* 5 (1985): 9–48.

17. Francesco De Sanctis, Atti Parlamentari, Camera dei Deputati, *Discussioni*, January 23, 1874, 708.

18. Anna Maria Mozzoni, *La donna e i suoi rapporti sociali*, now in her *La liberazione della donna* (Milan: Mazzotta, 1975). On the legal status of Italian women, see Simonetta Soldani, "Prima della Repubblica: Le italiane e l'avventura della cittadinanza," in *La democrazia incompiuta: Donne e politica in Italia dall'Ottocento ai nostri giorni*, ed. Nadia Filippini and Anna Scattigno (Milan: Franco Angeli, 2007), 41–90.

19. The *scuola normale* was not truly a secondary school until 1896, when it became compulsory to have three years of postprimary schooling (*complementare*, technical or lower-level gymnasium) to attend. Vocational schooling depended on the Ministry of Agriculture, Industry and Trade (hereafter MAIC).

20. Cf. Silvia Franchini and Paola Puzzuoli, eds., *Gli istituti femminili di educazione e di istruzione (1861–1910)* (Rome: Archivio Centrale dello Stato, 2005). Also available at http://www.archivi.beniculturali.it/pubblicazioni-free.html (Accessed June 17, 2009).

21. Cristina Belgiojoso, "Della presente condizione delle donne e del loro avvenire," *Nuova Antologia* 1 (1866): 106–107.

22. Ministero della Pubblica Istruzione (hereinafter MPI), *Documenti sulla istruzione elementare nel regno d'Italia*, vol. 3 (Florence: Botta, 1870), 172.

23. MPI, *Documenti*, 3: 175–176; Simonetta Soldani, "S'emparer de l'avenir: les jeunes filles dans les écoles normales et les établissements secondaires de l'Italie unifiée," *Paedagogica Historica* 40, 1–2 (April 2004): 123–142.

24. *Relazione presentata alla Camera dei Deputati il 22 dicembre 1866 dal Ministro dell'istruzione pubblica Berti intorno ai sussidi distribuiti per l'istruzione elementare e per l'educazione popolare del regno* (Florence: Botta, 1867), 3.

25. "L'istruzione della donna," *La Rassegna settimanale*, May 26, 1878, 387.

26. Felix Pécaut, *Deux mois de mission en Italie* (Paris: Hachette, 1880), 270–271.

27. See Giuseppe Saredo, *Vicende legislative della Pubblica istruzione in Italia dall'anno 1859 al 1899* (Turin: Utet, 1901), 253–277, for the *Relazione* to the Italian Legislative Decree of 1885 proposing a system of girls' secondary education.

28. The development of these institutes is not yet well known; see Giancarlo Rocca, "Gli istituti religiosi femminili e l'educazione delle donne in Italia tra Otto e Novecento," *Seminarium* 1–2 (2004): 209–258.

29. In 1892 female enrolments counted 1,167 and 583 respectively: cf. MAIC, Direzione Generale della Statistica, *Istruzione secondaria e superiore, convitti maschili e femminili, anno scolastico 1889–90* (Rome 1892), and *BUMPI*, April 12, 1893 and May 10, 1893.

30. Maria Bobba, "Gli studi della donna," in *La Donna Italiana descritta da scrittrici italiane* (Florence: Civelli, 1890), 373, 368.

31. Anna Maria Sorge, "L'evoluzione dell'istruzione normale e la documentazione conservata nell'Archivio centrale dello Stato," in *L'istruzione normale dalla legge Casati all'età giolittiana*, ed. Carmela Covato and Anna Maria Sorge (Rome: Archivio Centrale dello Stato, 1994), 51–54.

32. *Il Ministero dell'Istruzione pubblica e la educazione del popolo e della donna in Italia* (Rome: Cecchini, 1900), 98.

33. Cf. MPI, *Relazione presentata a S.E. il ministro della pubblica istruzione prof. Nicolò Gallo sugli Istituti femminili di Educazione e di Istruzione in Italia* (Rome: Cecchini, 1900). Private and religious institutes numbered 1,429; of these, 543 had postprimary courses modelled on the *Normali* or on the *Scuole superiori femminili*.

34. *Il Ministero dell'Istruzione pubblica*, 94.

35. These and the following numbers include the girls enrolled at the state and "*pareggiate*" schools, those run by local authorities or religious orders that granted recognized diplomas. In addition, 621 girls were enrolled at music and art academies.

36. In 1900 there were only ten all-girl technical schools (in Milan, Turin, Padua, Rome, and Palermo) and around ten "additional classes" reserved for girls in northern cities. Moreover, there were two gymnasia with all-girl classes in Milan and Rome (soon to be imitated in Girgenti, Trani, and Vicenza), and two female sections at technical institutes in Milan, soon to be joined by two more in Turin and Palermo.

37. Laura Torretta, "A proposito del Ginnasio Femminile di Milano," *La Corrente*, June 5, 1904.

38. Ibid.

39. Compare the French reform launched in 1902, which divided the *lycées* into two branches, in name at least almost equal: see Antoine Prost, *Histoire de l'enseignement en France 1800–1967* (Paris: Colin, 1968), 254–258.

40. The 1910 *Annuario Ecclesiastico* spoke of around 200,000 female pupils (aged between six and eighteen) enrolled at religious institutes: Riccardo Truffi, "Per l'istruzione della donna in Italia," *Annali della Istruzione Media*, June 1927, 532.

41. Agenda voted by the first congress on Female Practical Activities held in Milan in May 1908, quoted in Alessandrina Gariboldi, *L'Educazione e l'Istruzione della donna in Italia* (Reggio Emilia: Cooperativa tipografi, 1909), 30.

42. Giulio Fasella, *La riforma della scuola media e l'istruzione della donna* (Milan: n.p., 1906), 7.

43. "L'Educazione nuova e le conferenze di Sofia Bisi Albini," *Vita femminile italiana* 5 (1908): 546–547.

44. Ufficio Centrale di Statistica, *Gli Istituti per la Istruzione media e normale e la loro popolazione scolastica negli anni 1911–12 e 1916–17* (Rome: Provveditorato Generale dello Stato, 1921).

45. Françoise Mayeur, *L'enseignement secondaire des jeunes filles sous la Troisième République* (Paris: Presses de la Fondation Nationale des Sciences politiques, 1977).

46. MAIC, *Ordinamento dell'istruzione industriale e commerciale in Italia* (Rome: Bertero, 1911), 3. On this type of school, see also Aldo Tonelli, *L'istruzione tecnica e professionale di Stato nelle strutture e nei programmi da Casati ai giorni nostri* (Milan: Giuffré, 1964).

47. Giovanni Calò, *Il problema della coeducazione e altri studi pedagogici* (Rome: Dante Alighieri, 1914), 87.

48. Giovanni Gentile, *Esiste una scuola in Italia? Lettera aperta al ministro della pubblica istruzione on. Berenini* (May 4, 1918), republished in his *Il problema scolastico del dopoguerra* (Naples: Ricciardi, 1919), 8.

49. *Annuario Statistico Italiano* 1919–1921. At that date there were also 33,600 pupils (7,200 girls and 26,400 boys) enrolled in the corresponding courses at private institutes: *Annuario della Pubblica Istruzione*, 1924.

50. Law no. 1176 of July 17, 1919, article 7, excluded women from only those roles "that imply public jurisdictional powers or exercising political rights and power, or concern the military defence of the state."

51. See table 4. 2 this volume. For a well-balanced judgement of the reform, see Gabriele Turi, *Giovanni Gentile. Una biografia* (Florence: Giunti, 1995), 316–337, and Jurgen Charnitzky, *Fascismo e scuola. La politica scolastica del regime, 1922–1943* (Florence: La Nuova Italia, 1996 [German original, 1994]). In 1925 women teachers at secondary schools had fallen to 4,723, but they still made up over one-third of the total (*Annali della istruzione media*, 5 (1927): 455–475).

52. Giovanni Gentile, "La politica scolastica del regime" (1929), now in Gentile, *La riforma della scuola* (Florence: Le Lettere, 2003), 341.

53. Truffi, "Per l'istruzione della donna," 531.

54. *Annali della pubblica istruzione*, 1924.

55. On the consequences of the reform and the trends that it began, see Monica Galfré, *Una riforma alla prova: La scuola media di Gentile e il fascismo* (Milan: Franco Angeli, 2000).

56. In 1925, there were 12,783 girls enrolled at *ginnasi* and 3,261 at *licei*; their numbers grew very little in the following years: *Annali della pubblica istruzione*, 1929.

57. Charnitzky, *Fascismo e scuola*, 419.

58. *Statistica dell'insegnamento medio per l'anno scolastico 1936–37* (Rome, 1940) and *Annuario Statistico Italiano*, 1941. On the growing favor shown toward employment in the tertiary sector, see Victoria De Grazia, *How Fascism Ruled Women: Italy, 1922–1945* (Berkeley: University of California Press, 1992), chapter 5.

59. *Le Scuole per la Donna* (Rome: Ministero dell'educazione nazionale, 1940), 4–5.

60. In the state *Istituti magistrali* female enrolment increased from 23,500 in 1931 to 72,500 in 1939, but this represented only 66 percent of the total, confirming the efficacy of the steps taken by the Fascist regime to reintroduce men into teacher training: *Annuario Statistico Italiano*, 1934 and 1941.

61. The girls at the gymnasia leapt in the same years from 15,000 to 42,500, from 27 to 36 percent.

62. Giuseppe Sangiorgio, "Il nuovo regolamento dei concorsi-esami di Stato per le cattedre d'insegnamento medio e per l'abilitazione all'esercizio professionale," *Annali della Istruzione Media* 4 (1926): 255–256.

63. The pass rate among girls for the school leaving and qualification exams was constantly superior to the boys in every type of school: *Annuario Statistico Italiano*, 1941.

64. Tina Tomasi, *La scuola che ho vissuto* (Livorno: La Nuova Fortezza, 1987); and Marcella Olschki, *Terza liceo 1939* (Palermo: Sellerio, 1993).

65. For a description of the so-called *Carta della Scuola*, see Charnitzky, *Fascismo e Scuola*, 454–469.

66. In 1939, girls were 37 percent of pupils at the lower secondary schools and 38.5 percent at the upper schools, even though illiteracy among women fell very slowly.

67. MPI, *La ricostruzione della scuola italiana* (Rome: Centro Didattico Nazionale, 1950).

68. School of Barbiana, *Lettera a una professoressa* (Florence: LEF 1967), 16. The school was run by the parish priest of Barbiana, Don Lorenzo Milani.

69. In 1983–1984, girls accounted for 49.5 percent of the total, and 53.6 percent in the 14–18 female age group, against 52.9 percent in the male one, as recalled by Marcello Dei, "Cambiamento senza riforma: la scuola secondaria superiore," in Soldani and Turi, *Fare gli italiani*, 2: 87–127.
70. Marcello Dei, "Lo sviluppo della scolarità femminile in Italia," *Polis* 1 (April 1987): 143–158 and Giorgio Franchi, Barbara Mapelli, Vito Librando, "Donne a scuola. Scolarizzazione e processi di crescita di identità femminile negli anni '70 e '80," in *Donne a scuola in Europa: Lo studio di alcuni casi europei di interventi formativi al femminile*, ed. Lea Battistoni (Milan: Franco Angeli, 1989), 31–51.

Chapter 5

Between Modernization and
Conservatism: Spain*

Consuelo Flecha

In Spain the secondary educational system developed somewhat differently than else-where in Western Europe, in part because the Reformations and the Enlightenment had less of an impact. During the eighteenth century, secondary education and university study were tightly intertwined, leaving few opportunities for women. Additionally, Catholicism and Catholic institutions heavily marked the educational system, further limiting the possibility for girls to pursue secondary education, despite periods of more liberal reform. In 1900 the illiteracy rate among the female population over the age of ten was 66 percent, more than 20 percentage points higher than the male rate of 45.7 percent.[1] This figure testifies to the backwardness of the Spanish school system in general; as a result, only a small minority of girls benefitted from secondary education until the recent past. This reality understandably has conditioned scholarship on the subject, which remains, on the whole, underdeveloped.

Historians of education have written extensively about the emergence of the Spanish educational system during the nineteenth century, treating such subjects as State initiatives and reforms, the presence of Church schools, the social origins of pupils and teachers, and the characteristics of academic disciplines or textbooks. Alongside monographs, there now exist general overviews that include the pioneering works of Antonio Viñao, Federico Sanz, and Emilio Díaz.[2] A number of collective volumes usefully complete these overviews, notably those coordinated by Manuel de Puelles and by Maria Nieves Gómez,[3] and a special issue of the journal *Historia de la Educación*.[4] These contributions point to the troubled history surrounding the conception of secondary education as it moved from being primarily preparation for university studies to becoming the prolongation of primary education. They also highlight the early coexistence of both classical humanities and a more modern curriculum that aimed to prepare young men for their civil responsibilities. These studies, however, include little information about the secondary education of girls, who for most of the nineteenth century were considered ineligible for more advanced education. Nor do these studies reveal much about the history of the lay and religious schools that offered girls some education.

Given the weight of the Catholic Church in Spanish political and educational history, many historians have examined issues of secularization and traced how religious issues determined governmental policy with respect to education.[5] The rise of anticlericalism in the nineteenth century paradoxically increased the demand for denominational schools, as statistics concerning boys' secondary education and girls' primary education clearly reveal.[6] Women's historians, for their part, have explored such issues as the place of women within the system of social relations created by the first liberals,[7] the influence of discourse and ideologies on female identity,[8] as well as the weight of Catholicism and its associations.[9] These studies all highlight the ways contemporaries, be they liberal or conservative, believed female behavior should uphold the established social order.

Scholarship on girls' education has emphasized the weight of conservative educational models when schools began to develop. These models had a strong domestic focus, ensuring the permanence of gendered spheres between men and women.[10] Within this social order, however, the figure of the mother educator opened opportunities for women to lay claims for a better education. Although feminism as a general social movement only developed in the twentieth century, historian Mary Nash has argued that "education became the core agenda in first-wave feminism," and isolated voices from the eighteenth and nineteenth centuries argued for its role in cultural transformation and female emancipation.[11] In the second half of the nineteenth century, reforms in secondary and higher education generated debates that included women, as work on higher education and middle-class women's efforts to enter hitherto exclusively male professions has shown.[12] More recently, scholarship has examined how the conservatism of Franco's dictatorship affected girls' education.[13] Still, no synthetic study of girls' education exists as of yet, despite the dynamism of women's history in Spain.

Debating Girls' Education (1700–1867)

Spanish Enlightenment discourse opened a space for both men and women to articulate a call for instruction as the prerequisite for happiness and progress. Writers argued for the importance of acquiring useful learning to improve industrial production and daily life. The Enlightened despot, King Charles III, who ruled between 1749 and 1788, actively promoted the development of schools and included girls in his efforts. Besides creating free schools for girls in Madrid,[14] he also encouraged the opening of *Colegios femeninos* (girls' colleges). In 1785 he authorized the first woman in Spain, María Isidra de Guzmán (1768–1803), to pass the doctoral exam at the University of Alcalá for her work in philosophy. This young woman from the nobility was known for her knowledge and erudition and the monarch wished to reward her in public.[15] He also supported the entrance of women into the private associations known as *Sociedades Económicas de Amigos del País* (Economic Societies of Friends of the Country), which were created to stimulate economic and intellectual development.

The Aragonese writer and translator Josefa Amar y Borbón (1749–1833) was among the women to enter these economic societies. A physician's daughter, she possessed an uncommon level of education for a woman of her period and quickly achieved notoriety for her defense of the intellectual capacity of women and their role in society. She argued that women should be treated with equality and included a vibrant defense of women's education in two late eighteenth-century writings that have become foundational texts for Spanish feminism. In addition to reading and writing, she defended women's right to study history, arithmetic, Latin, modern languages, and geography. In one of her works, she wrote: "If women were to receive the same education as men they would do as much or more than them." Several years later she added: "Instruction is beneficial for all; and women should not be exempt from its benefits because of the ways it allows them to switch occupations, and make retirement more gratifying... It is said, and rightly so, that in both sexes there is equality of talent."[16]

Her voice and those of others were not taken into account, however, in the liberal constitution of 1812 (also known as the constitution of Cadiz), which laid the groundwork for the creation of a national system of education by proclaiming that every village should have a primary school. While liberals advocated universal education, in reality this meant boys' education. In the years between 1812 and the passage of the Moyano law in 1857, which gave definitive shape to the public educational system, boys' education received far more attention than that of girls, and coeducation was prohibited.

Despite a public commitment to education, the disastrous financial situation of the State administration left the terrain open to private educational institutions, particularly for the education of girls. Both lay teachers and religious congregations opened *Colegios de Señoritas* (Ladies' Colleges) for the daughters of the bourgeoisie. True, the teaching congregations faced specific obstacles during periods when progressive anticlericals were in power, notably with the prohibition of new admissions to religious orders and then the disentailing of Church lands from 1836 to 1855, which affected cost-free education in many schools. Still the Concordat of 1851 between the Catholic Church and the Spanish State represented the triumph of a socially conservative elite, and the 1857 Moyano law allowed religious orders to maintain houses for education and teaching, if they complied with requirements, which included allowing inspections by the authorities. Most important, religious teachers could claim exemption from the qualifications required for lay teachers in private secondary education.

During the first half of the century, young women could only pursue an education beyond primary education in private institutions known as *Casas de Enseñanza* (teaching houses), which offered courses in history, geography, literature, French, as well as the female accomplishments: drawing, embroidery, and piano. Female teachers also gave classes in pupils' houses in cities throughout Spain, but particularly in the larger ones such as Madrid and Barcelona. Both English and French women earned a living in Spain teaching the daughters of the wealthy.[17] The French woman Cornelia Sesment was headmistress of a school in Ronda (Málaga), which featured common classes in grammar, geometry or geography, and optional classes in sewing, embroidery or music. Alongside these lay women, religious congregations, such as the *Salesas*, the *Compañía de María*, the *Carmelitas de la Caridad* or the *Madres Escolapias* offered programs that went beyond the primary level.[18]

The educational reforms of the 1830s and 1840s that led to a clear distinction between secondary and university education also saw the emergence in 1838 of *Escuelas Normales* (Normal Schools) for boys only, as well as higher primary schools in 1845. The existence of higher primary schools stimulated a demand for more professional training among women. In 1847 the regional government of Navarre inaugurated the first *Escuela Normal* for girls, an initiative that soon spread to other provinces.[19] When Madrid founded its *Escuela Normal Central de Maestras* (Central Normal School for Female Teachers) in 1858, the authorities explicitly emphasized their concern to encourage improvements in teaching methods.

These schools attracted young women aspiring to be teachers, but also those who wished to pursue studies in a serious fashion; for many decades, they offered a rare possibility for secondary-level education, particularly in provincial areas. The study plan was oriented toward the exercise of a profession that was considered socially ideal for women, especially for single women, as their salary gave them personal independence. As a result, even middle-class women who did not plan to work as teachers obtained teaching diplomas in these schools,[20] and some then went on to work.

Other middle-class girls attended drawing and embroidery classes in special patriotic schools where teaching was organized by the *Junta de Damas de las Sociedades Económicas de Amigos del País* (Ladies' Committee of the Economic Societies of the Friends of the Country). These committees encouraged women to put their talents to work in the production of clothing; sewing the national dress could provide an honest way to make a living in several Spanish provinces.[21] Other young women studied music at the *Conservatorios* (music academies) and art at the *Escuelas de Pintura, Escultura y Grabado* (Schools of Painting, Sculpture and Engraving). The Fine Arts Academy at Vitoria admitted its first female students in 1840; one of these students, Daría Imbert, went on to give classes at the Academy in 1866–1867.[22]

Girls were left outside the scope of the secondary school system as it acquired shape in the 1840s. The intermediate stage of schooling that emerged in these years reflected a progressive, meritocratic, and profoundly masculine vision. The *bachillerato* (baccalaureate) that crowned these studies was "aimed at the upper and middle classes, that is, to those who are more active and entrepreneurial...to those who legislate and govern; to those who write, invent, direct and provide society with impetus, leading it along the different paths of civilization; in short, to those who are the souls of nations, who move people and cause happiness or misfortune."[23] The Law of Public Instruction of 1857 did not explicitly change these objectives; as a result, the task of opening up these studies to other social groups, including women, still remained.

Reforms and New Initiatives in Girls' Secondary Education (1868–1900)

In the six years following the September Revolution (1868–1874), a political period that commenced with the expulsion of Isabella II, there were few changes in the general approach to female instruction despite attempts to introduce profound changes in all areas of society, including education. Feminist activity in other countries received coverage in the press, but there was scarcely any activity in Spain. As a result, the initiatives of this period came mainly from individuals who opened more vocationally oriented institutions, or from girls themselves who forced open the doors of the male *Instituto de Segunda Enseñanza* (Institutes of Secondary Education). At the same time, older forms of girls' schooling also developed and became more serious as expectations about feminine intellectual standards rose.

The Rector of the University of Madrid, Fernando de Castro (1814–1874), played an important role in what historian Rosa Capel has described as "the opening of women's cultural horizons."[24] In 1869 he organized a series of conferences for women that paved the way for the creation of the *Escuela de Institutrices* (School of Governesses), which started its activity in 1869, and then one year later for the *Asociación para la Enseñanza de la Mujer* (Association for the Teaching of Women). The latter adopted the organization of the German *Lette Verein*, established in 1866 in Berlin.[25] The *Escuela de Institutrices* offered quality education and its aim was to "provide young women with the elements considered most important in women's intellectual, moral and social culture and prepare those who are to dedicate themselves to teaching and education."[26] The concern to train governesses testifies to their common presence in upper-middle-class Spanish families.

The *Asociación para la Enseñanza de la Mujer* was particularly active, encouraging the creation of more vocationally oriented institutions for women throughout the provinces. It was set up in Vitoria in 1879, Málaga in 1886, Valencia in 1888, Granada in 1889, and some years later in Mallorca, Barcelona, Bilbao, Seville, and Zaragoza; its principle concern remained "encouraging the education and instruction of women in all spheres and conditions of social life."[27] Along similar lines, the *Escuela de Comercio para Señoras* (Ladies' School of Commerce) opened courses in 1878 in Madrid for women aiming to work in trade; similar courses were then opened in Valencia (1884), and a few years later in Mallorca. The *Escuela de Correos y Telégrafos* (The Post and Telegraph School) started classes in Madrid in 1883.

These efforts to raise the cultural level of women from the middle classes were not intended to distance them from the bourgeois feminine ideal, despite the accent placed on the acquisition of knowledge with practical applications. The statutes of the *Asociación para la Enseñanza de la Mujer* emphasized the concern to prepare women "to fulfill correctly the duties imposed by their sex . . . and so that they may apply their activity to different professions."[28] This orientation reflected the influence of German philosopher Karl Friedrich Krause on the founder of the Association. Kraussist thought emphasized the transformation of people and institutions through an ethical renewal of social and political life that required new educational initiatives.[29] The initiatives of the *Asociación para la Enseñanza de la Mujer* also enjoyed the support of men from the *Institución Libre de Enseñanza* (Free Institute of Teaching) who encouraged these programs that prepared women for a variety of remunerated employment without neglecting training in their domestic roles.[30]

The First Girls in Secondary Education Institutes

The six-year period following the September Revolution of 1868 also witnessed the first efforts on the part of young women to obtain access to the *Institutos de Segunda*

Enseñanza. Two decrees in October 1868 had proclaimed greater freedom of teaching, thus reorganizing secondary and university teaching without any reference to the presence of women. As legislators did not see this question as a priority, they did not deem it necessary to exclude women explicitly. A grammatical loophole—in which the masculine is used when two words of different gender coincide—in the legal guidelines that only included girls explicitly at the primary school level encouraged the first girls to seek entrance into the baccalaureate program of the *Institutos*, the most prestigious and demanding level of education prior to university. Unlike elite schooling elsewhere in Western Europe, however, the *Institutos* were less focused on the classical humanities, and the receipt of the *bachillerato* did not depend on class attendance. Indeed the State organized three types of matriculation, which allowed students to enroll while studying with private tutors or in a private *collegio*. This flexibility opened opportunities for young women in the final quarter of the century.

In 1870, the *Instituto* of Huelva in Andalucía was the first to receive an application from thirteen-year-old Antonia Arrobas. The director's correspondence with ministerial authorities raised two issues: "the inconveniences of bringing young boys and girls together in the same class or having to build twice the number of *institutos* so that they may study with the correct separation." The Minister decided that the girl "has the right to that for which she applies."[31] When the same situation arose a few months later in Barcelona, officials emphasized the need to avoid "the inconvenience that could be caused by bringing together both sexes in the classroom," and stated that girls had to study under a regime of private teaching and only attend the *Instituto* to sit their exams. Legislation did not prohibit coeducation but the heads of *Institutos* were warned about the problems it could provoke.

Despite these conditions, a total of 162 students took exams between 1870 and 1881, testimony to a new concern among middle-class families to have their daughters pursue secondary studies.[32] Reports from individual *Institutos* refer to the admission of girls and their academic performance. In Valencia, the first occasion was reported thus: "Attention should be drawn to the fact that nine young ladies enrolled and obtained very good marks…, the excellent application shown by all is worthy of applause, and particularly for one […] who came out on top of all the students."[33]

Still, many expressed concern about the dangers of coeducation and having boys and girls study the same academic subjects. Educational reformer Pedro de Alcántara García, a teacher from the *Escuela Normal Central de Maestras* and member of the *Asociación para la Enseñanza de la Mujer*, argued, "The complications deepen with secondary education for women, where haste has meant that they are offered the same type as is offered to men. If in relation to the latter, it does not meet, by a long shot, the goals for which it was designed, imagine what will happen with the women." He went on to warn that "secondary education for women is not a preparation or authorization to undertake another order of studies, nor courses that are taken with the exclusive aim of obtaining a diploma that will open the door to certain university degrees." Rather this education for the feminine sex represented "culture or general education."[34]

Opinions such as these were widespread in these years and help explain that in 1881, following consultation, ministerial authorities agreed with the Council of Public Instruction to create institutes for girls, although the first ones did not appear until 1929. A report by the Public Instruction Council for the Minister of Public Works in 1882 stated: "It would seem advisable to create, at least in some areas, centers for instruction where women could acquire the knowledge taught in secondary education, with a practical orientation, and even higher education in sciences and letters with that same orientation." Minister of Public Works Germán Gamazo used this argument in a Royal Order of 1883, which stated the need for "legislative reforms to provide a teaching plan for women."[35]

Reformers such as Pedro de Alcántara García supported the idea of separate schools for girls where they would study "as little as possible abstract and detailed subjects that, far from interesting them, harm and embarrass their intelligence . . . Nothing of the *bachillerato* or wisdom." He recommended the spread "over as much of the country as possible, in cities and villages, of what we could call, let us say, *Feminine Institutes*."[36] Similar to discussions elsewhere in Western Europe, Spanish reformers argued that girls should avoid studying disciplines designed for those destined to create, direct, and drive society, such as law, industry, trade, or agriculture.

This concern for respecting gender differences undoubtedly prompted the Royal Order of 1882, which prohibited new admissions of women to university, eliminating one of the most important reasons for pursuing the *bachillerato*. As a result, at the start of the 1882–1883 academic year it was also stipulated that admissions would not be accepted for "enrolments in courses of secondary education for young ladies," as it was understood to be of no use to them. This regulation was short-lived and girls were readmitted to secondary education one year later.[37]

The number of girls studying for the *bachillerato* grew steadily in the *Institutos* up to the end of the century; of these, 50 percent completed the courses and obtained the baccalaureate diploma. A significant number had studied previously at the *Escuela Normal*, or would do so after taking some subjects at an *Instituto*. Given the opposition to coeducation, it often was easier for them to attend the former in order to acquire the knowledge required for the diploma.

Personal accounts reveal that relationships between boy and girl students were not always easy, as in some *Institutos* boys rejected girls' presence in the classroom.[38] However, girls seeking to pursue their studies did receive support from some, notably the well-known Spanish feminist Concepción Arenal (1820–1893). At the Hispanic-Portuguese-American Pedagogical Congress held in Madrid in 1892, she stated: "The first thing a woman needs to do is affirm her personality, regardless of her status, and recognize that, single, married or widowed, she has duties to fulfill, rights to claim, dignity that depends on no one else, a job to do, and she must realize that life is a serious, solemn thing, and if she treats it like a game she will undoubtedly become the toy."[39] This was an explicit and authoritative message from a woman who had undertaken public responsibilities and enjoyed great recognition as a lawyer, prison reformer, philanthropist, and writer.

Alongside the relatively few public *Institutos* that welcomed girls, a large number of Catholic schools catered to the daughters of Catholic families for whom this more traditional form of schooling held more appeal.[40] Nineteen foreign religious teaching congregations set up almost forty schools for girls in Spain alongside the older existing schools. Most of these came from France following the introduction of a secular public education system between 1881 and 1904.[41]

The *Escuelas Normales* also underwent reforms in the final decades of the century. In particular, they responded to the growing demand for education among girls who did not want to experience coeducation with boys in *Institutos*. The 1882 reform of the *Escuela Central Normal de Maestras* established an entrance requirement (pupils had to have completed higher primary education), increased the program to four years, and added subjects offered at the School of Governesses as well as new ones, such as industrial painting and languages.[42] These changes attracted pupils from a wider social sphere and testified to the ways postprimary schooling for girls had begun to hold more appeal even among conservative families.

Expanding Opportunities in the New Century (1900–1936)

The first two decades of this period were marked by continually changing liberal and conservative governments, and while attempts were made to introduce reforms the lack of political stability thwarted their implementation. The dictatorship of Primo de Rivera (1923–1930) saw a rise in social discontent and unrest. With the advent in 1931 of the Second Republic, the government announced an ambitious program to transform the country, but these plans were cut short by electoral defeat in 1933. Despite the tumultuous politics of these years, and a general failure of the more ambitious plans to modernize and democratize secondary education, women saw a significant expansion in their educational and occupational opportunities.

While the new century brought fewer reforms than the previous one, they were more effective and led to an improvement in the way schools worked. A Ministry of Public Instruction and Fine Arts was created in 1900 in an attempt to address the problems associated with a highly deficient school network, particularly in rural areas. Girls still suffered from high rates of illiteracy, 71 percent compared to 56 percent for boys,[43] and were expected to help their mothers in their homes. At the same time, the Ministry recognized the need to promote postprimary education in the context of social and economic development. Despite growth in the number of both male and female secondary pupils, their overall percentage remained very low (and far below the rates in other West European countries).[44]

Nonetheless, there were signs of new expectations among women with respect to schooling. Increasingly girls requested the right not just to sit for exams, but also to become fully enrolled secondary students sharing the classrooms of *Institutos* with their male peers, despite the perception this was unseemly feminine behavior.[45] Measures were therefore taken in the early years of the twentieth century to ensure respect for morality while they were at school; areas were set aside for them, they were always accompanied, and in class they were seated separately from boys.

By 1910 most *Institutos* had some experience of girl students, although resistance to their presence continued, notably among teachers who remained attached to traditional gender roles. At the *Instituto* of Ciudad Real in 1907, pupils had to listen to the following opening address given by one of the school's teachers:

> Allow us with all seriousness and respect to this solemn event, to fulfill a higher act of gallantry, for the ladies and mistresses who, like delicate *bouquets of orchids* and flowers in a living room, represent the most artistic decoration of these seats: allow us to bid farewell with flowers to those whom we have received with flowers. You, when facing your future, do not impose your will by the strength of reason but by the empire of your suggestion, and face strength with meekness, reflection with feeling, and contrariness with your charms... Receive the worship of our admiration and respect, as letters, arms and gallantry, have always been the blazoned shield of every Spaniard.[46]

An abyss separated this speaker from a more modern world in which a growing number of young women wished to live, including those listening to him.

Two legislative orders marked the beginning of a new era recognizing women's rights that had hitherto been disputed. The first, on March 8, 1910, freed women from having to "consult Superiors" when they wished to enroll for university, as "the general sense of the legislation for Public Instruction is not to distinguish on the basis of sex, authorizing the admission of male and female students on equal terms"; that of September 2 of the same year stated that "the possession of different academic qualifications will enable women to undertake all professions related to the Ministry of Public Instruction."[47]

These orders reveal changing attitudes toward women's place in society, since they granted women admission to the university, along with the right to work as professionals at *institutos*, universities and libraries, archives and museums. Women quickly took advantage of this;[48] by April 1911, the drawing teacher at the *Instituto* of Valencia was a woman, Julia Gomis Llopis, who had studied at the Royal Academy of Fine Arts of Saint Carlos. There were now more reasons for obtaining the *bachillerato* and going on to university and this was reflected by a rapid increase in the number of female students: they grew by 230 percent between 1910 and 1920 so that by 1920 girls constituted 8.4 percent of all baccalaureate students, and 19.2 percent of all secondary students (baccalaureate, teacher training and professional tracks).[49]

However, unease about boys and girls sharing the same classrooms did not wane.[50] The testimony of a student at Madrid's *Instituto* of San Isidro in 1920–1921 highlights boys' growing awareness of their female peers:

> Attending the *Instituto*, along with us, are a growing number of *bachilleras* [female students], and because of their sex and out of respect, the teachers of the different subjects seat the girls on benches separated from the boys and usually close to the teacher. But this year, undoubtedly due to the feminist doctrines that are rife, the number of girls has grown considerably and the places on the benches in the chemistry classroom were insufficient.[51]

This feminine presence stimulated by feminist demands also disturbed teachers as they made clear: "I understand that the boys who come here to study and prepare their future may get up to some mischief, but you, my ladies, who are only here for leisure, pay attention and be quiet."[52] Statements such as these reveal the obstacles girls encountered in imagining a future along similar lines to their brothers.

New Private and Religious Schools for Spanish Girls

Opportunities for girls also expanded greatly in the private sector with the creation of both religious and lay institutions catering to the growth in demand among female pupils wishing to study for the *bachillerato* in a more congenial atmosphere. One such school, founded by the American Protestant missionary Alice Gordon Gulick, was the *Instituto Internacional*. Here American teachers taught girls in a cultured and pluralistic atmosphere, but also enabled some to take the baccalaureate exams at the *Instituto*.[53] In 1918, Catalina Vives (1895–1979), the first woman to obtain a doctorate in natural sciences in Spain, founded the *Liceo Feminino* (Lycée for Girls) in Madrid, where women teachers taught subjects for the *bachillerato* and offered teacher training.[54]

In liberal Catholic circles, the pedagogue Pedro Poveda (1874–1936), founder of the Catholic Association *Institución Teresiana*, was among the most active promoting more rigorous standards for boys and girls alike. He created *Academias Teresianas* (Academies of Saint Teresa) in several Spanish cities starting in 1911 with women teachers who taught and prepared girls for the exams at the *Escuelas Normales* and the *Institutos*.[55] The first female university residence was opened in Madrid in 1914, and in 1923 the *Instituto Católico Femenino* (Catholic Institute for Girls) opened to teach the baccalaureate.[56]

The existing religious *colegios*, while more conservative in their social ethos, also responded to the growing expectations middle-class families had that their daughters would pass exams and enter the university. By 1926 the following *colegios* prepared girls for *Instituto* exams: *Compañía de María, Carmelitas de la Caridad, Compañía de Santa Teresa de Jesús, Esclavas del Sagrado Corazón, Sagrado Corazón de Jesús, Dominicas de la Presentación* and *Ursulinas*.[57] As a result, by 1930, of the 35,717 students enrolled in universities, 6.3 percent were women.[58]

Alongside traditional secondary schools, a number of new creations highlight the new demands among some sectors of the lower middle and middle classes. One enterprising center was the Iberoamerican Center for Women's Popular Culture founded in Madrid in the early twentieth century, offering teaching grouped into four sections: Industry; Science, Letters, and Arts; School of Commerce; Schools of Mothers of Families.[59] Interest in more rigorous training in domestic economy also prompted the creation in 1911 of the *Escuela del Hogar y Profesional de la Mujer* (Home and Professional School for Women) in Madrid. The founding statutes insisted that the young girl "should possess some elements of human knowledge to facilitate the learning of various professions," but also argued that "it is necessary for the woman to learn and know how to care for her children and the home."[60] This education continued to assign women to the family and the domestic setting, while providing training for "feminine" professions: governess, midwife, nurse, stenographer, and typist. Similarly, in 1909, within the circle of Catalan Catholic feminism,

Francesca Bonnemaison (Barcelona, 1872–1949) promoted the Institute of Culture and Popular Library for Women in Barcelona; two years later it began offering secondary, professional, domestic education, and languages. The older Asociación *para la Enseñanza de la Mujer* (Association for the Teaching of Women) created in 1869 also offered secondary teaching from 1919 on. This expanding access to education did not seek, however, to challenge the social, political, and economic relations between men and women.

Educational Reform and National Regeneration

Although the 1920s were marked in Spain by the dictatorship of Primo de Rivera (1923–1930), the period was also characterized by a considerable expansion in schooling as a result of rising demand, and both State and municipal investments. For girls, reforms in the *bachillerato*, notably the emergence in 1926 of a three-year elementary *bachillerato* in smaller provincial cities, as well as the creation of girls' *Institutos* (beginning in 1929) made secondary schooling more easily available.

The decision to create the latter built on earlier attempts. In 1910 a girls' *Instituto* was attached to the existing *Instituto* in Barcelona. Its objective "was to broaden primary teaching, through general culture, following the official study plan of the *bachillerato*." Since it was for girls, this training was completed "with specific, practical and intuitive subjects," particularly sewing, house craft, cooking and domestic economy, and basic health care, presumably, so that they would not forget what their condition as women required of them.[61] This first initiative of public secondary education, which allowed girls to study without sharing classrooms with boys, attracted families hostile to coeducation: "families that retained their daughters at home, despite their love for progress and feminine culture, have hurried to take them to the new *Instituto*, whose defining characteristic lies in teaching the fairer sex separately."[62]

Anxiety about coeducation was evident throughout Western Europe in the second half of the nineteenth century,[63] and in the 1920s a series of International Congresses on Secondary Education devoted careful reflection to the subject (in 1922, 1924, and 1928).[64] In Spain, the debate about coeducation continued to rage as some families renounced secondary studies for their daughters because schools were mixed, while others considered single-sex schools to be out of date after so many years of sharing classrooms. This international context combined with dictator Primo de Rivera's concern to develop secondary education helps to explain the creation of two girls' *Institutos*, one in Madrid, the *Infanta Beatriz*,[65] and another in Barcelona, the *Infanta Cristina*, to teach the elementary, not university, *bachillerato*. The government justified these measures by arguing:

> Secondary education, as a way of obtaining a general culture, of stimulating vocations, and of preparing for other higher levels of learning [...] affects the great majority of the middle class; due to the steady increase in female attendance, it constitutes an ongoing matter of interest to large sections of public opinion and requires preferential attention from the government.[66]

In Barcelona the creation of the *Infanta Cristina* did not mean closing the already existing section, neither did it lead to the closure in Madrid of the Institute-School created in 1918, with a section for girls studying the *bachillerato*. However, the girls' *Institutos* were short-lived because the first government of the Second Republic transformed them into coeducational schools in 1931.

The government's encouragement of girls' secondary education reflected a new reality. It was no longer seen as a "mere preparation for Faculty studies." Increasingly, "a large number of the young ladies who attend *Institutos*"[67] did so with the intention of pursuing mid-level professions. And the elementary *bachillerato* responded well to this demand, leading as it did to careers in nursing and midwifery, which were taken up predominantly by women.

Teacher training opportunities for young women increased as well with the growth of primary schools for girls and the resulting opportunities this offered for women teachers. Two reforms of the *Escuelas Normales* illustrate the new importance that was given to this training. The first in 1914 strengthened the cultural and pedagogical content of this training; the second in 1931, following the Republican victory, introduced decisive innovations, as it required the *bachillerato* for entrance, introduced male teachers alongside their female peers, and established a study plan based on pedagogical training.

The Franco Era: Educational Conservatism and Girls' Determination

The educational policies of the Franco regime sought to turn the tide of feminine aspirations for equal access to secondary and higher education. Political discourse in particular emphasized mothers' roles in rebuilding the nation demographically and morally, and educational initiatives drew heavily on religious values to direct girls toward this designated social mission.[68] The ideology of the early Francoist regime was not coherent, however, as the Catholic Church, the Falange (Spain's fascist party), and the Traditionalists all jockeyed for influence. So while conservatism certainly characterized policies toward girls' education, these were imbued at times with specifically political values; women were expected to contribute to national regeneration.

A number of specific measures encapsulate the new educational conservatism. During the Civil War (1936–1939), in the area under Franco's control, girls' *Institutos* returned, thanks to a decree in September 1936 prohibiting coeducation and establishing separate schools for boys and girls; the decree remained in effect until 1970.[69] Then the reorganization of the *bachillerato* in 1938 instituted a single seven-year classical program of study aimed at teaching "the classical and humanistic studies of our sixteenth century [. . .] of our imperial age, to which the heroic vocation of our youth is returning."[70] Associated with a reform of the internal regime of *Institutos*, these initiatives sought to make secondary education an incubator for elites, with women resolutely sidelined into their domestic role. The General Director of Secondary and University Education incarnated these attitudes when he stated in 1940: "An attempt should be made to channel the great stream of girl students away from the feminist pedantry of becoming female *bachilleras* and university students; these must be the exception [...our goal must be] to direct them towards their own magnificent feminine being, which should develop at home."[71] Women students studying for the classical *bachillerato* were obliged to follow a complement of courses in sewing and domestic economy, the so-called *enseñanzas del hogar* (education for the home), that previously had only been taught in the girls' section of the *Instituto de Bachillerato* in Barcelona and at Madrid's *Instituto-Escuela*.

By the 1950s, however, the educational system was in serious need of modernization with 30 percent of school-age children still unschooled. In secondary education, a reform in 1953 divided the *bachillerato* into cycles: after four years, students could obtain an elementary *bachillerato*, two more years led to a superior *bachillerato*, and a final year corresponded to preuniversity study. This measure encouraged student enrollments and the number of girls increased steadily: 38.8 percent in 1959–1960, 45.2 in 1969–1970, and 48.8 in 1975 at the end of Franco's regime.[72] These students came mainly from urban areas, however, as there were still few *Institutos* in rural areas and as the dominant private Catholic sector did not prepare girls for the *bachillerato*.[73]

The General Education Law in 1970 decisively improved opportunities for public secondary education for girls. Coeducational schools reemerged in a political, economic, and cultural context where feminist and democratic ideals encouraged girls to distance themselves from the domestic model that had weighed so heavily on Spanish ideals of femininity.[74] As a result, in the last third of the twentieth century the number of girl students rose rapidly, along with their awareness of their right to enjoy the benefits of these studies and to seize initiatives with respect to their professional future.

Conclusion

Paradoxically, in Spain the absence of public measures in favor of girls' secondary education encouraged a number of girls and young women to pursue their quest for more serious education alongside boys or in the lay or religious institutions that emerged to fill a void. The fact that boys' *Institutos* were neither numerous—there were fifty-eight in all of Spain in 1875 and ninety-six in 1930—nor particularly rigorous undoubtedly facilitated girls' access to this level of education in a country where the spread of primary education was very slow even as late as the first third of the twentieth century. Girls took advantage of reforms, such as those of 1926 and 1953, to attend *Institutos* and persevered in the face of conservative measures, such as those in 1938. This tendency was strengthened by the Education Law in 1970, so that by 1976–1977, girls represented 50 percent of all secondary students, a trend that continues today (part of secondary education was made compulsory in 1990). Spain's exceptionally high rates of female enrolments in secondary schools and university in Europe are testimony to a growing awareness among women of what education can achieve. Particular emphasis has been laid in recent years in Spain on public policy toward equal opportunities and measures to make equality between men and women a guiding principle in all fields, and these initiatives have succeeded in changing mentalities, behaviors, and expectations.

Notes

*Translated by Patrick Partridge.
 1. Antonio Viñao, "Historia de un largo proceso," *Cuadernos de Pedagogía* 179 (1990): 47.
 2. Antonio Viñao, *Política y Educación en los orígenes de la España contemporánea* (Madrid: Siglo XXI, 1982); Federico Sanz, *La Segunda Enseñanza Oficial en el siglo XIX* (Madrid: Ministerio Educación Ciencia, 1985); Emilio Díaz, *Evolución y desarrollo de la Enseñanza Media en España. 1875–1930* (Madrid: Ministerio de Educación, 1988).
 3. Manuel de Puelles, ed., *Política, legislación e instituciones en la educación secundaria* (Barcelona: ICE, 1996); María Nieves Gómez, ed. *Pasado, Presente y Futuro de la Educación Secundaria en España* (Seville: Kronos, 1996).
 4. *Historia de la educación. Revista Interuniversitaria* 17 (1998): 5–178.
 5. Julio de la Cueva and Feliciano Montero, eds. *La secularización conflictiva. España 1898–1931* (Madrid: Biblioteca Nueva, 2007).
 6. Teódulo García, *La polémica sobre la secularización de la enseñanza en España (1902–1914)* (Madrid: Fundación Santa María, 1985), and José Álvarez, *Mater Dolorosa, la idea de España en el siglo XIX* (Madrid: Taurus, 2003).
 7. Gloria Espigado, "Las mujeres en el nuevo marco político," in *Historia de las mujeres en España y América Latina*, ed. Isabel Morant (Madrid: Cátedra, 2006) III, 27–60.
 8. Dolores Ramos-Teresa Vera, eds. *Discursos, realidades, utopías* (Barcelona: Anthropos, 2002).
 9. Julio de la Cueva and Ángel-Luis López, eds. *Clericalismo y asociacionismo católico en España* (Cuenca: Eds. Universidad Castilla-La Mancha, 2005); Manuel Revuelta, *La Iglesia española en el siglo XIX.* (Madrid: Universidad Comillas, 2005).
10. Pilar Ballarín, *La educación de las mujeres en la España contemporánea* (Madrid: Síntesis, 2001), and Irene Palacio, "Mujeres aleccionando a mujeres. Discursos sobre la maternidad en el siglo XIX," *Historia de la Educación* 26 (2007): 111–141.
11. Mary Nash, "The Rise of the Women's Movement in Nineteenth-century Spain," in *Women's Emancipation Movements in the 19th century: A European Perspective*, ed. Sylvia Paletschek and Bianka Pietrow-Ennker (Stanford: Stanford University Press, 2004), 246.
12. Geraldine Scanlon, *La polémica feminista en la España contemporánea (1868–1974)* (Madrid: Akal, 1986); and Consuelo Flecha, "Women at Spanish Universities. The Origin of Their Presence," in *New Women of Spain*, ed. Elisabeth de Sotelo (Münster: Lit Verlag, 2005), 397–409.
13. Anny Brooksbank Jones, *Women in Contemporary Spain* (Manchester: Manchester University, 1997); Aurora G. Morcillo, *True Catholic Womanhood. Gender Ideology in Franco's Spain* (De Kalb: Northern Illinois University Press, 2000); Victoria Lorée Enders and Pamela-Beth Radcliff, eds. *Constructing Spanish Womanhood: Female Identity In Modern Spain* (New York: State University of New York Press, 1999).
14. Paloma Pernil, *Carlos III y la creación de las escuelas gratuitas de Madrid* (Madrid: UNED, 1989).
15. Vicente de la Fuente, *Historia de las universidades, Colegios y demás establecimientos de enseñanza en España* IV (Madrid: Viuda de Fuentenebro, 1889), 125–129.
16. Josefa Amar y Borbón, "Discurso en defensa del talento de las mujeres y de su aptitud para el gobierno." [1786] Memorial Literario VIII, no. 32 (1876): 415. Victoria López-Cordón, ed., *Discurso sobre la educación física y moral de las mujeres* [1790] (Madrid: Cátedra, 1994), 170–171.
17. Carmen Simón, *La enseñanza privada seglar en Madrid (1820–1868)* (Madrid: Instituto Estudios Madrileños, 1972), 127–133.
18. Isabel de Azcárate, *El Monasterio de la Enseñanza de Barcelona, 1645–1876* (Barcelona: PPU, 1993); Pilar Foz, *Fuentes primarias para la historia de la educación de la mujer en Europa y América: Compañía de María, 1607–1921* (Roma: Tipografía Vaticana, 1989); Isabel Florido, *Acción educativa de las Hijas de la Caridad en España (1783–1893)* (Granada: Publs. Hijas Caridad San Vicente de Paúl, 1988).

19. Esther Guibert, *Historia de la Escuela Normal de Navarra* (Pamplona: Príncipe de Viana, 1983).
20. Between 1856 and 1860, women obtained 3,519 elementary teaching diplomas and 324 secondary teaching Diplomas. *Anuario Estadístico*. http://www.ine.es/inebaseweb/pdfDispacher. do?td=25777&ext=.pdf
21. Estrella de Diego, "La educación artística en el siglo XIX," in *El trabajo de las mujeres, siglos XVI–XX*, ed. María-Jesús Vara & Virginia Maquieira (Madrid: Universidad Autónoma, 1996), 480–488.
22. Francisca Vives, "La enseñanza artística de la mujer en Vitoria en el siglo XIX," *Sancho el Sabio* 8 (1998): 213–220.
23. Antonio Gil de Zárate, *De la instrucción pública en España*, 3 vols. (Madrid: Colegio Sordo-Mudos, 1855) II, 1 [ed. fascimile Madrid: Pentalfa, 1995].
24. Rosa-María Capel, "La apertura del horizonte cultural femenino," in *Mujer y Sociedad en España (1700–1975)* (Madrid: Ministerio Cultura, 1986), 109–145.
25. Raquel Vázquez, *La Institución Libre de Enseñanza y la educación de la mujer* (Betanzos: Lugami, 2001), 35.
26. Asociación para la Enseñanza de la Mujer, *Reglamento Escuela de Institutrices* (Madrid, 1882), art. 1º.
27. *Estatutos de la Asociación* (Madrid, 1882), art. 1º.
28. Ibid.
29. Morcillo, *True Catholic Womanhood,* 10.
30. Raquel Vázquez, *La Institución Libre*, 23–84. Elvira Ontañón, *Un estudio sobre la Institución Libre de Enseñanza y la mujer* (Valencia: Universidad Valencia, 2003).
31. Decree of 27 May 1871. See Antonio Correa, "Las primeras alumnas en el Instituto de Segunda Enseñanza de Huelva," in Gomez, *Pasado, Presente y Futuro de la Educación*, 132–133.
32. Report of 1882 on "Women who have completed studies in universities and official institutes in Spain in recent years," in *Textos y Documentos sobre Educación de las Mujeres*, ed. Consuelo Flecha (Seville: Kronos, 1997), 235–259.
33. Instituto Provincial de Valencia, *Memoria del curso de 1878 a 1879*, 10.
34. Pedro de Alcántara García, "Caracteres, sentido y dirección de la educación fundamental de la mujer," *Revista de España* 104 (May–June 1885): 210, 213.
35. Consuelo Flecha, *Las primeras universitarias*, 90.
36. Pedro de Alcántara García, "Caracteres, sentido y dirección," 227.
37. Consuelo Flecha, *Las primeras universitarias*, 89–91.
38. María-Carmen Oña, "Digno de Memoria," *La Universidad* I, 12 (February 1888): 90. "Lo del Instituto Provincial," *La Universidad Española* 2, 72 (November 1889): 240. Carmen de Zulueta and Alicia Moreno, *Ni convento ni college. La Residencia de Señoritas* (Madrid: Residencia-Estudiantes, 1993), 82.
39. Concepción Arenal, "La educación de la mujer," in *La emancipación de la mujer en España*, ed Mauro Armino (Madrid: Júcar, 1974), 67.
40. Manuel Revuelta, *La Iglesia española en el siglo XIX*, 144.
41. Antonio Molero, "Influencias europeas en el laicismo escolar," *Historia de la Educación* 24 (2005): 164–165.
42. Carmen Colmenar, *Historia de la Escuela Normal Central de Maestras* (Madrid: Universidad Complutense, 1988).
43. Mary Nash, "The Rise of the Women's Movement," 255.
44. Emilio Díaz mentions that in 1874 there were 17 secondary students per 10,000 inhabitants, and by 1930 this had only risen to 27. *Evolución y Desarrollo de la Enseñanza Media en España. 1875–1930* (Madrid: CIDE, 1988), 508.
45. Consuelo Flecha, "La incorporación de las mujeres a los Institutos de Segunda Enseñanza en España. 1870–1910," *Historia de la Educación* 17 (1998): 159–178.
46. Instituto General y Técnico de Ciudad-Real, *Memoria Curso 1907–1908* (Ciudad-Real: Enrique Pérez, 1908), 5–6.
47. Consuelo Flecha, *Las primeras universitarias*, 93.
48. Consuelo Flecha, "Profesoras y alumnas en Institutos de Segunda Enseñanza (1910–1940)," *Revista de Educación*, nº extr. (2000): 269–294.
49. Rosa-María Capel, *El trabajo y la educación de la mujer en España (1900–1930)* (Madrid: Ministerio de Cultura, 1986), 415, 420.

50. Antonio Viñao, "Espacios masculinos, espacios femeninos. El acceso de la mujer al bachillerato," in *Mujer y educación en España*, ed. Coloquio de Historia de la Educación (Santiago: Universidad, 1990), 567–577.
51. José Gavira, "Diario de un estudiante del Instituto de San Isidro (1920–1921)," *Anales Instituto Estudios Madrileños* 9 (1973): 531–532.
52. Ibid., 541.
53. Carmen de Zulueta, *Misioneras, feministas, educadoras. Historia del Instituto Internacional* (Madrid: Castalia, 1984); Morcillo, *True Catholic Womanhood*, 15.
54. La Dirección, "Liceo femenino en Madrid," *La Escuela Moderna* XVIII, 326 (1918): 711–714.
55. Between 1911 and 1919 eleven boarding-school academies were opened: in Oviedo (1911), Linares (1912), Jaén (1913), Madrid (1914), Málaga (1914), La Carolina (1915), León (1917), Barcelona and Teruel (1918), Ávila and Burgos (1919). In the following decade they were opened in San Sebastián, Alicante, Bilbao, Córdoba, Sevilla, Santander and Santiago de Compostela. Capel, *El trabajo y la educación*, 354.
56. Aurora G. Morcillo, *True Catholic Womanhood*, 130–140.
57. Félix Restrepo, "Labor de las Asociaciones católicas en la educación y en la enseñanza," *Razón y Fe* (1926): 282.
58. Anuario Estadístico. http://www.ine.es/inebaseweb/pdfDispacher.do?td=150212&ext=.pdf
59. "Centro Iberoamericano de Cultura Popular Femenina," *Unión Iberoamericana* XXI, 1 (1907): 20.
60. Real-Decreto December 7, 1911, en *Gaceta de Madrid (día 22)*, 356, 698–700.
61. *Anuario Estadístico ciudad de Barcelona* (Barcelona: Heinrich) 15 (1916): 237–238; 16 (1917): 222.
62. Juan Caballero, "Instituto de Barcelona para la Segunda Enseñanza de la Mujer," *La Escuela Moderna* 238 (1911): 407, 414–424.
63. See chapter 12 by James Albisetti in this volume.
64. Consuelo Flecha, "La educación de las mujeres después del 98," in *La educación en España a examen (1898–1998)*, ed. Julio Ruiz (Zaragoza: Mec-Diputación, 1999), II, 333–335.
65. Natividad Araque, "El Instituto Femenino Infanta Beatriz (1900–1930)," *Revista Complutense de Educación* 12, no. 2 (2001): 753–781.
66. *Historia de la Educación en España* (Madrid: Ministerio Educación, 1982), III, 218.
67. Ibid., 219.
68. Morcillo, *True Catholic Womanhood*, 46–76.
69. *Historia de la Educación en España* (Madrid: Ministerio Educación, 1991), IV, 296–297.
70. Ibid., 303.
71. José Pemartín, *Qué es lo nuevo* (Madrid: Espasa-Calpe, 1940), 142.
72. *Anuarios Estadísticos*. http://www.ine.es/inebaseweb/libros.do?tntp=25687.
73. Isabel Grana, "Las mujeres y la segunda enseñanza durante el franquismo," *Historia de la Educación* 26 (2007): 262–265.
74. Marina Núñez and María-José Rebollo, "La prensa femenina de post-guerra: Materiales para la construcción identitaria de la mujer española," in *Etnohistoria de la Escuela* (Burgos: Universidad de Burgos, 2003), 231–246.

Chapter 6

Toward the Recognition of Their Educational Rights: Portugal

Helena C. Araújo, Cristina Rocha, and Laura Fonseca

In Portugal the weight of women's destiny as housewives and mothers was stronger than in other countries, according to Ana de Castro Osório, a remarkable republican feminist of the early 1900s. She attributed what she considered to be a sad state of affairs to the atmosphere of "moral asphyxia" and the flawed nature of girls' convent education that contributed to the "imprisonment" of women.[1] For bourgeois and aristocratic girls, the domestic orientation of girls' education, so harshly condemned by Osório, prevailed from the eighteenth to the early twentieth century in home education, private institutions, and state schooling. At the end of the nineteenth century, however, feminist voices began calling for changes in girls' education, citing, in particular, examples from Switzerland, Germany, and especially America.[2] Nonetheless, more egalitarian opportunities for girls in secondary education only developed in the second half of the twentieth century and primarily after the 1970s.

No synthetic study of girls' secondary education in Portugal exists as of yet; as a result, this chapter presents a broad historical outline of the pathways of girls' "secondary" education, showing how institutional offerings and pedagogical programs changed, most notably thanks to the impact of women's demands. The arguments developed draw on scholarship that remains patchy, given a long tradition of defining secondary education within a male historical narrative. Girl students appear in studies of the Portuguese high school but often only through a discussion of rising female enrollments; hence specific gender constraints and limitations are not the object of these studies.[3] Nonetheless, recent work from scholars in women's history has begun to reveal the variety of settings within which girls received an education that went beyond elementary skills. In particular, studies now highlight the importance of the Ursulines and other religious orders in the growth of girls' secondary education, although much work remains to be done on nineteenth-century convent education.[4] Other scholars have revealed the role of governesses in the nineteenth century and notably the significant presence of foreign governesses teaching a wide range of subjects within a home setting. The life stories of pioneering lay women teachers in the Republican period and the early years of military dictatorship between 1910 and the 1930s have highlighted the emancipating effect of formal education and work in the public sphere despite the weight of gender norms.[5]

On the whole, the twentieth century is far better known, due to the combined efforts of historians and sociologists. As a result, studies exist that take into account the impact of pressure from middle-class as well as upper-working-class girls in the achievement of a form of "mitigated meritocracy" in secondary vocational education.[6] More recently, Teresa Pinto has studied the tensions within girls' vocational education as it developed at the end of the nineteenth century, as training for industry and trade confronted the ideal of domesticity for working-class women.[7] With regard to secondary education, Cristina Rocha has examined the conservative 1930s when strong State regulation in both public and private schooling contributed to growing elitism.[8] The work of Irene Pimentel explores how the political and ideological project of forming women to become "angels of the home" was pursued during the *Estado Novo* (1933–1968). She highlights, in particular, how the Catholic Church and the State joined forces to control women's destiny both within the family and formal schooling.[9] Other studies have explored the educational strategies of parents during the same period, notably the attraction private Catholic institutions held for Lisbon elites.[10] For the second part of the twentieth century, analysis of statistical data has established a chronology of girls' education, building on earlier studies that showed the increase in attendance of girls' secondary schools following the revolution of April 25, 1974 and the ensuing democratization of the Portuguese educational system.[11] Nonetheless, despite the growth of studies taking girls' education as their primary focus, much work remains to be done.

Limited Educational Offerings for Girls (1700–1890)

State intervention to create an educational system in Portugal began in the eighteenth century. This effort developed after the expulsion of the Jesuits in 1759 that until then had provided the predominant model of secondary education for boys. For historian Antonio Nóvoa, this represented an attempt "to Europeanize Portugal," through secularization and the strengthening of the power of the nation-state.[12] Between 1759 and 1772–1773, the State established classes in Greek, Latin, Rhetoric, and so on, known as *aulas régias*,[13] as well as primary schools, specialized courses in trade and industry, and the College for the Nobles (*Colégio dos Nobres*).[14] Alongside these State-sponsored initiatives, religious orders and private teachers continued to teach the sons of wealthy families. Boys were, however, the privileged recipients of both private and public offerings. Girls only gained access to state primary schools under Queen Maria I in 1815, learning Christian doctrine and needlework in addition to reading, writing, and mathematics.

Despite limited opportunities for women, the eighteenth century "pedagogical revolution" generated arguments in support of women's education.[15] The foreign-educated intellectuals (*estrangeirados*) Luís António Verney[16] (1713–1792) and Ribeiro Sanches[17] (1699–1783) were the most prominent Enlightenment figures to include a discussion of girls' education within the framework of a general reform of mores and mentality. They were particularly critical of Jesuit education and in their writings provided guidelines for the reforming actions of the absolutist prime minister, Pombal (1750–1777).

Enlightenment Discourses

Adopting an epistolary form, Verney's work *Verdadeiro Método de Estudar* (1742) (The True Method of Study) focuses mainly on what we now recognize as secondary education. Letter XVI presents a plan of study for girls inspired by the writings of Fénelon and Rollin. Verney argued that men and women had equal abilities and thus women should be given opportunities to improve their minds. He proposed courses that would contribute to the physical and intellectual betterment of girls and would enable them to perform appropriately as married women. These courses included: religion, Portuguese, grammar, arithmetic, geography and sacred history, Greek and Roman History, the history of Portugal, domestic economy, needlework, singing, music and dancing; Latin was reserved for nuns and noble girls.[18] Like the French pedagogues who inspired him, Verney emphasized women's role as mothers and gave his course a domestic orientation.

Ribeiro Sanches published his *Cartas sobre a Educação da Mocidade* (Letters on the Education of Youth) in 1760 immediately following the 1759 Reform Act. A critic of the inherited privileges of the nobility, he argued for the need to provide aristocratic girls with a serious education, in order to prevent vice and ignorance as mothers.[19] He also sought inspiration in French examples, such as the Maison Royale de Saint-Cyr founded by Madame de Maintenon in Versailles. Ribeiro urged the transformation of nuns' convents into girls' schools. Girls' courses should counteract morally dangerous leisure activities such as those associated with "reading love novels, [and] poems, not all of them sacred [...]."[20] His proposals also included the daughters of the bourgeoisie. Their education should prepare them for three possible roles in society: as respectable mothers, nuns, or single women. In his proposals, moral improvement and an education for domesticity went hand in hand. But he also envisioned that this education could allow single women to find paid work outside the home and live independent lives as governesses or private teachers. Both authors believed in the separate roles of men and women and tailored their educational aims to these roles; men were destined to govern the nation, and women to govern the home. For women this involved responsibility for raising children and running the domestic economy. Clearly they assumed that the domestic sphere was propaedeutic to the public sphere for their male children.

Convent Education in the Eighteenth and Nineteenth Centuries

In the second half of the eighteenth century, a number of religious orders began to offer schooling for girls.[21] The Italian order of *Salesianas* (S. Francisco de Sales) founded the *Mosteiro da Visitação* in 1782, in Lisbon, and opened a boarding school for daughters of the aristocracy. In addition to the rudiments, the curriculum included lessons in Portuguese, French, Italian, English, and Latin grammar. Girls also studied geography, the harpsichord, and needlework. This program prepared students for salon society as well as life at the royal court, without neglecting the religious and moral lessons necessary to maintain pious conduct.[22]

The Ursulines were among the most active of religious orders in girls' education, starting schools for girls in 1753 in Pereira (near Coimbra) and later in Viana (1777) and Braga (1784). Their rulebook emphasized that education was the key to sanctity; hence girls' education was useful for their own redemption, but also for the church and the State, since "everything depends on the good education of mothers." The first woman director of the school in Pereira, Leocádia Vaía, argued for the importance of a moral Christian education: "A woman should neither be a slave, nor a libertine."[23] The school prospered in the nineteenth century, thanks to the support of political authorities, even during periods of intense anticlericalism, and it acquired a reputation for the education of elite girls. Beginning in 1789, the Ursulines were forced to receive poor girls also, in order to receive a state subsidy. The boarders, however, were strictly separated from the poor. The former followed a curriculum that bore close resemblance to that offered in the Mosteiro da Visitação. In the last decade of the nineteenth century, it included lessons in the "art of receiving in society" as well as the feminine accomplishments: music, singing, poetry.[24]

Education in the Home

Throughout the nineteenth and the early decades of the twentieth century, wealthy families hired governesses (mostly foreign women) to teach their daughters at home. Lessons typically included the study of the Portuguese language, foreign languages (French, German, and English), history, geography, arithmetic, geometry, and the piano. An 1899 unpublished diary, written by a middle-class seventeen-year-old girl, offers detailed information on the number of English, French, and German lessons she received from foreign governesses. Her diary also describes studying the piano with members of her family of musicians, as well as learning to sew.[25]

The republican and feminist pedagogue and novelist, Alice Pestana (1860–1929)—writing under the pseudonym Caiel—was one among many girls from her middle-class origins who received schooling at home with foreign governesses.[26] In Caiel's novels, there are constant references to foreign governesses who taught girls what was considered suitable to their social standing; many of her female characters are from middle- or upper-middle-class backgrounds. Germana, in *Desgarrada* (1902), was taught French, English, piano, singing, embroidery, and decorative painting (see also *Madame Renan*, 1896). In many other Portuguese novels, the English governess figures prominently in these social environments. In *Os Maias* (1888), one of the most famous novels of the realist author Eça de Queiroz (1848–1900), an English governess is in charge of Maria Eduarda's daughter and follows the family everywhere. These novelistic representations emphasize how girls' education in these social groups was directed at ensuring a proper marriage. However, some of the literature clearly shows that the governess' teachings also promoted the concept of the "new woman" who was not so much a "lady of leisure," but someone who, in physical terms, would be well-fitted to bear offspring, and in cultural terms would be a suitable partner for her husband and a good manager at home. Such a representation can be found in the education of the daughters of Feliciana Montanha, the widow of a rich Northern landowner in *Genoveva Montanha* (1897). In some novels, authors deploy national stereotypes, depicting the English governess as "ugly," and "raw-boned," wearing unfashionable flat boots and sandals. At other times the foreign governess is admired for her character and ideals that are favorably contrasted with the unhealthy ways of living of Portuguese society.[27]

Education in the Colonies

Within Portuguese colonies, religious orders were primarily responsible for opening schools for girls. In the eighteenth and nineteenth centuries in Brazil, convent schools for girls were established in Rio (1739, *Recolhimento das Órfãs da Misericórdia*) and Baía (1716). As in Portugal, the goal was to prepare girls for marriage. An *educandarium* (a religious college) was established at Macaúbas (1727) in the Minas Gerais and became famous in the nineteenth century for elite girls' education.[28]

From the end of the nineteenth century, and more intensively in the first half of the twentieth century, eighteen religious orders, including the *Franciscanas Hospitaleiras Portuguesas* and the French *Congregação S. José de Cluny*,[29] established schools for both white and indigenous girls in Portugal's African, Asian, and Melanesian colonies. Secondary schools open mainly to white girls were more common in Angola and Mozambique. In the latter colony, in particular, the Portuguese State encouraged the Catholic Church and its missions to play a central role in teaching the black population, thus furthering its project of a colonial empire.[30]

Calls for Change in Girls' Education: Expanding Opportunities (1890s–1930s)

Debates about women's role and gender destiny increased notably toward the end of the nineteenth century, carried in large part by feminists and pedagogues who were influenced by models of girls' education encountered in Europe and

North America.[31] In many places Auguste Comte's scientific model of industrial development underpinned positivist republican perspectives encouraging a vision of domesticated womanhood. Women were entrusted with the care of the household as a haven against the competitiveness and aggressiveness of the public world, using their "weapons of feelings" to help men find a solution to their dilemmas. Scientists, politicians, and moral economists justified the unique role of women as mothers, housekeepers, and wives on the basis of their "natural" and biological characteristics. Science was believed to have demonstrated biologically, physiologically, and sociologically that women's true role was maternity.

Debating and Comparing Models of Girls' Education

In Portugal, a small feminist movement emerged toward the end of the century in the context of crisis within the Monarchy. Primarily influenced by feminists in North America and England, the main actors in this movement shared republican and anticlerical sympathies and sought to reconcile these ideologies with feminist objectives. For feminists such as Ana Castro Osório or the republican freethinker and teacher Maria Velleda (1871–1955), the right to education and to work was an essential precondition for greater equality between men and women.[32] This message, however, was hard to hear in a predominantly Catholic country. For Catholics, women's mission was within the home, a belief that the Popes' encyclicals reinforced through their insistence on women's subordination to the family.[33] The Roman Catholic Church argued strongly that women should dedicate their lives to the well-being of their families, and it condemned female emancipation and work outside the home.

Within this context of political opposition and debate, feminists and educators looked to more progressive societies for models. The *Partido Progressista* (Progressive Party) within the monarchist and parliamentarian government responded to calls for change in girls' education by sending personalities such as the woman educator, novelist, and feminist, Caiel to visit and write reports on institutions throughout Europe where girls pursued secondary education. Caiel expressed her admiration in 1892 for the splendor, organization, and structure of Girton College in Cambridge and Queen's College in London, of the *École Normale de Sèvres* in France and the *École Supérieure des Jeunes Filles* in Lausanne, and of Vassar College in the United States.[34] The writer, politician, secondary teacher, and Freemason, Borges Graínha (1862–1925) also wrote a report on secondary education of both girls and boys. Over a decade after Caiel, he described visiting the *École Ménagère* and *École Secondaire et Supérieure des Jeunes Filles* in Geneva, the *Lycée* for girls in Schaerbeek near Brussels, as well as other schools in France and Italy. Graínha focused on girls' schools in Belgium and Switzerland, small countries like Portugal, which nonetheless offered advanced schooling unlike his home country. In the context of projected school reforms, he argued that the future girls' secondary school should aim to achieve an integral education, combining both intellectual and domestic training, thus responding to women's needs in civil society.[35]

Private and Public Secondary Schooling for Girls

A number of religious orders established girls' *colégios* (secondary private schools) at the end of the nineteenth century. These orders included the Doroteias (1866, Lisbon; 1879, Porto), the Franciscans from Calais (1875, Aveiro), and the Dominicans who opened the *Colégio S. José*.[36] Secular institutions also spread in the late 1890s. In Porto, for example, the *Colégio Inglez para Meninas* (English Private School for Girls) as well as the *Colégio das Inglesinhas* opened for the existing British colony.[37] Alongside the conservative Catholic institutions, well-known educators in Porto also founded institutions, such as the *Colégio Moderno*, the *Colégio Nossa Senhora da Estrela*,[38] and the *Colégio Araújo Lima*, where coeducation was in place before the turn of the century. These schools took inspiration from local educators and pedagogues, such as João Diogo (1868–1923), following the *Escola Nova* (inspired by Adolphe Ferrière) or Carolina Michaelis (1851–1925), an influential intellectual and university professor who supported equal education for girls.

Formal state secondary education for girls emerged at the end of the nineteenth century, fifty years after the first public high school for boys (1836), with the opening of the *Escola Maria Pia*. The boys' *liceu* was modeled partially on the French *lycée* and the German *Gymnasium*, and was similarly selective in their social body. Formed to educate and recruit the governing elite, they proposed a humanistic education and also opened the doors to university education.[39] Although a law adopted in June 1880 allowed girls to enter boys' schools as day students, few girls took advantage of this possibility until the late 1880s when debates about girls' education became more public.[40] As a result, female enrolments in secondary schools initially grew very slowly, as table 6.1 shows.

The *Escola Maria Pia* was founded with the support of feminist groups who adhered to its liberal, republican, and masonic agenda, and started its activities in the academic year of 1885/86. A three-year general course was followed by one-year specialized courses on trade, typography, telegraphy, or needlework. The study plan included: Portuguese, French, English, German (optional), arithmetic, geometry, trigonometry, natural history, geography, history, physics, "chemistry and hygiene," drawing, calligraphy, needlework, "morals and law," and "women's duties." In 1888 the vocational orientation was dropped due to the lack of demand. Although the local authorities saw it as an *école primaire superieure*, the school was actually a secondary school where almost all the subjects from the *liceu* were taught by secondary school teachers.

Table 6.1 Girls' and boys' enrolments in boys' high schools, 1892–1901[a]

Years	Girls	Boys
1892/93	26	3591
1900/01	66	3406

[a] Nóvoa and Santa-Clara, *Liceus*, 28.

In 1906 this institution became the first girls' *liceu* and changed its name to *Liceu* Maria Pia. The course program reflected the influence of the French model, with an initial three-year cycle, followed by an optional two years. In the first three years, girls studied the same school subjects as boys but with fewer contact hours in each subject, in order to accommodate specific feminine subjects (see table 6.2).

Unlike secondary programs for girls throughout Europe, Portuguese pupils studied more physics and natural sciences than elsewhere.[41] The influence of Spencer's theories can be seen in the legislator's concern to avoid forming "*précieuses ridicules*," but rather competent mother-educators, thanks to their practical studies.[42] In the 1st cycle, 23 percent of contact hours involved studying "subjects for the female sex," reflecting the assumption that girls should also be prepared for their domestic functions as mothers, spouses, and housekeepers. Indeed the introduction to the 1906 law stressed:

> Domestic education is a necessary complement for the general instruction of women. Hygiene is central for housekeepers, for mothers; cooking needs to be studied in parallel with chemistry, with physiology, since this art needs all the sciences for its execution;

Table 6.2 Weekly contact hours for the 1st and 2nd cycles of the *liceu* for both sexes, 1906–1918

Subjects/School Years	1st Year		2nd Year		3rd Year		4th Year		5th Year	
	M	F	M	F	M	F	M	F	M	F
Portuguese Language and Literature	5	4	4	3	3	3	3	3	3	2
Latin (optional for girls)							3	3	3	3
French	4	3	3	3	3	3	2	2	2	2
English or German	—	—	4	3	4	3	3	3	3	3
Geography and History	3	2	3	2	2	2	2	2	2	2
Physics and Natural Sciences	3	2	2	2	4	3	4	3	4	4
Mathematics	5	4	4	3	4	3	3	3	3	3
Drawing	3	2	3	2	3	2	3	2	3	2
Physical Education	3	2	3	2	3	2	3	2	3	2
Total (partial)	26	19	26	20	26	21	26	23	26	23
Subjects limited to girls										
Moral Education, Economy, Hygiene and Cooking		2		2		2		1		1
Pedagogy								2		2
Music		2		2		2		1		1
Needlework		2		2		2		2		2
TOTAL	26	25	26	26	26	27	26	29	26	29

Source: Cristina Rocha and Manuela Ferreira, *As Mulheres e a Cidadania—as mulheres e o trabalho* (Lisbon: Horizonte, 2006), 126. (The table has been adapted).

domestic science by itself constitutes a marvelous scientific synthesis. Moral education, notions of common law, pedagogy, music need to be taught. Without them, a
family mother's education cannot be completed.

However, marriage was not the only destiny envisioned for adult women, since the
law also mentioned the possibility of work for women outside the home as an "honest way of gaining one's daily bread," for instance, as school teachers.[43] The second
cycle was quite similar to the first cycle regarding subjects and contact hours. Latin
was taught, although optional for girls, probably given the expectation that girls
were not necessarily applying for jobs or that they were not competing with boys for
the same jobs.

Women's Teacher Training Colleges

The expansion of schooling for girls stimulated a demand for women teachers.
With the Educational Act of 1878, four single-sex teacher training schools (*Escolas
Normais*) were created (two for each sex) in Lisbon and Porto and ten mixed-sex
schools opened in the provinces. As early as 1885, inspectors began to deplore the
low number of male students in teacher colleges. The creation of mixed-sex colleges
reflected a social demand: increasing numbers of women were attracted to teaching
as "a good future for women."[44]

Teacher training colleges included specific subjects considered suitable for
women: instead of "agriculture," women students learned "gardening"; instead of
the "duties of the citizen," the "duties of mothers in the family," instead of "rural,
industrial, and commercial economy," "home economics"; instead of learning "commercial writing," there was "needlework and embroidery." Academic subjects—such
as arithmetic, geometry, geography, notions of physics, chemistry, and natural history, drawing, pedagogy—took up less time in women students' timetables in order
to accommodate the hours devoted to needlework, embroidery, and applied drawing (six hours per week over the three years). Educators and politicians stressed that
women's education in teacher colleges should follow the gender models of the time.
Several reports emphasized the need to relate the subjects taught more directly to the
experience and needs of women's future lives. For instance, the 1889 Inspectorate
Report claimed there was not enough teaching of home economics and needlework
in the Lisbon women's teacher training college. Only some years later, in the 1901
Education Reform Act, did women's and men's curricula in teacher colleges become
more similar.

The Republican Period (1910–1926)

The Republic opened new opportunities for many women beyond education and
work. Some of the most restrictive measures concerning family life and the position of
women within it were abolished with the 1910, 1911, and 1918 republican laws. Women
acquired the right to act in legal matters. They were able to publish their own work

without their husbands' permission and no longer had to swear obedience to them. A wife was also able to request the suspension of paternal authority over children, when fathers put children in moral danger. Wives could not be compelled to return to their home if they left of their volition. It was also possible to ask for the investigation of illegitimate paternity, an innovative measure, generally nonexistent in other countries at the time. Paid maternity leave of two months was granted to working women in 1921, in the last stage of their pregnancy, regardless of whether they were married or not. And, finally, a divorce law was introduced, against the will of the Catholic Church and conservative sectors, allowing men and women to claim divorce on the same grounds.

In this context, the female *liceu* represented a clear break from the past and reflected the State's growing (albeit conflicted) recognition of women as human beings with *equal* rights. At the same time, the program for the Republican *new woman* proposed a framework that fit with considerations of what was thought to constitute a proper female education. At the outset of the Republic in 1910, girls represented 11 percent of the total number of state secondary pupils, a figure that rose to 26 percent by 1926.[45] The growth in girls' secondary enrolments was the product of both political and sociocultural forces. The Portuguese republican state consciously sought to modernize society and encouraged the instruction and education of women.

The *Liceu Maria Pia* did not charge fees in 1906, aiming to attract girls with few economic resources. However, in 1912, it became fee-paying, like the other *liceus*. New *liceus* emerged: the *Liceu Sampaio Bruno* in Porto in 1915/16 (which became the *Liceu Carolina Michaelis* in 1926) and the *Liceu Infanta D. Maria* in Coimbra in 1918–1919.[46] They started as girls' departments in male state high schools and later became autonomous girls' schools. Official documents in 1918 emphasized that the aim of secondary education for both sexes was the same; by then the specific feminine school subjects were no longer considered as core curriculum and were taught as supplementary subjects.

For girls from the lower classes, the development of higher primary schools, along the lines of the French *écoles primaires supérieures*, represented another opportunity to pursue postprimary education. These schools, created by law in 1911, only started to receive pupils in 1919. They aimed to bring together vocational knowledge and an academic preparation for pursuing studies after primary school. The promoters of these schools hoped they would help diminish social inequalities. Others suggested they served to shift the laboring classes' aspirations from the *liceu* to these types of schools.[47] Interestingly, girls were in the majority. Besides teaching industrial and agricultural studies, they taught home economics and trade, which may explain the presence of girls in great numbers. These schools were abolished, however, after the military coup of 1926.

Educational Politics from the Estado Novo (1933–1974) to the Present

Political turmoil in the years between the military coup of May 28, 1926 and the affirmation of an authoritarian State under Salazar in 1933 had repercussions on society as a whole. The "Estado Novo," from its birth, assured its own "stability."

New institutions were created, others were abolished, and yet others were put under strict conservative control. They constituted a repressive reply to a variety of political "disorders": strikes, demonstrations of the democratic opposition, and the resurgence of communist activities.

Women's status was particularly affected by the powerful influence of a conservative Catholic Church. The 1933 Constitution changed women's legal rights. The family was identified as a central institution and women were firmly anchored within it. Husbands recovered the legal power to force women to reside in the home. Among other setbacks, women lost the ability to do business, act as judges, or chief public executives; they were prevented from signing contracts, managing goods, or going abroad without the written consent of the husband.[48]

Under Salazar's long rule (1933–1968), educational policies emphasized the values of nationalism and authoritarianism. The State encouraged the preparation of elites within the *liceu*. Nóvoa has described educational policies in this period as *policies of restraint*, limiting the access to secondary state education. At the same time, policies supported the expansion of private secondary schools for the social elites.[49]

The new emphasis on women's special nature also produced an increasing opposition to coeducation, introduced during the former period of the Republic. The conservative right claimed that coeducation not only encouraged sexual promiscuity, but also constituted a strategy to demoralize the young generations. After the military coup of May 1926, this language intensified: critics accused coeducation of causing prostitution, encouraging free love, constituting a threat to the nation, and even bringing bolshevism to Portugal.

The support for coeducation at the end of the Republic came from a variety of radical educational and political traditions, including utopian socialists, feminists, and republicans in general. All saw schooling as a means of promoting greater equality for girls. Their arguments often centered on the "natural process" of coeducation since girls and boys were not separated in other activities of daily life. Thanks to coeducation, adolescents would lose their morbid curiosity concerning "sex matters." Other arguments emphasized the benefits for family life—boys and girls would acquire a better knowledge of the other sex—as well as the contribution to both boys' and girls' development: boys would lose their aggressiveness and their "Don Juan" behavior, while girls would gain in audacity and courage, adopting "a more virile way of life." Feminists argued that girls were as capable as boys and so coeducation would offer them the same rights. The *Conselho Nacional das Mulheres Portuguesas* (National Council of Portuguese Women) supported these arguments and criticized the military and authoritarian government when coeducation was abolished.

In practice the larger cities, Lisbon, Porto, and Coimbra, had state high schools exclusively for girls. In other towns, however, girls could attend the same *liceus* as boys either in separate rooms or separated in the same spaces: girls at the front, boys at the back. Boys and girls entered through different doors and were separated during recreation.

During this period, the girls' secondary school curriculum remained, nonetheless, the same as for boys (with the exception of needlework), following the reform of 1918 that extended the boys' curriculum to girls. However, stricter gender roles than in former Republican times were enforced through the hidden curriculum. In 1936, the *Mocidade Portuguesa Feminina* (Portuguese Female Youth) took over the teaching of non-compulsory subjects, "female" subjects (such as cooking, elementary nursing, floral decoration, etc.) under the guidance of the *Obra das Mães* (Mothers' Organization).

Girls in Secondary State and Private Schools

Despite the conservative nature of the regime in the 1930s, girls attended secondary education in greater numbers, mainly in private secondary schools (*colégios*). The State created new *liceus* for girls; the *Liceu Filipa de Lencastre* in Lisbon opened in 1928/29 and the *Liceu Rainha Santa Isabel* in Porto in 1933/34. Opportunities for postprimary schooling for girls also existed outside the *liceu*. The number of private secondary schools increased faster from the 1930s onward both for girls and boys. Some were boarding colleges, others enrolled only day students, and some had both regimes. They were costly and attended by the middle- and upper-middle classes (in trade, industry, the professions, and landowning).

As can be seen in table 6.3, girls' enrolments in private education expanded rapidly, representing more than half of the total girls' enrolments in the 1940s. These results were the product of policies that channeled the majority of pupils to private education and gave the Catholic Church the opportunity to establish many of its own schools.

Despite the growth of private institutions, girls also flocked into high schools and by 1960/61 outnumbered boys (table 6.4).

Table 6.3   Girls' secondary education: enrolment numbers and percentages[a]

Academic Years	State High Schools		Private Secondary Schools		Home Education		Total	
1920/21	2602	81.4%	237	7.4%	359	11.2%	3198	100%
1930/31	3321	73%	998	22%	232	5%	4551	100%
1940/41	5627	45%	6472	51.6%	433	3.4%	12532	100%

[a]*Anuários Estatísticos*, Lisbon, INE. Cristina Rocha, "Entre l'Enseignement Privé et l'Enseignement Public."

Table 6.4   Female and male pupils in *liceus* (high schools), 1930–1960[a]

Years	Female Pupils	Male Pupils
1935/36	6,007	11,875
1950/51	9,563	12,399
1960/61	23,177	22,883

[a]Nóvoa and Santa-Clara, *Liceus*, 28–29.

Silent Expansion from the 1950s Onward

From the 1950s to the 1960s, girls' secondary education expanded rapidly. This (silent) expansion marked a dramatic reversal of the earlier situation.[50] Laura Fonseca highlights the 1960s as a turning point, when girls began to outnumber boys in *liceus*, a period when a meritocratic discourse and human capital theory began to be heard. The 1960s appear as a period marked by the growth in both public and private secondary education.

The revolution of April 25, 1974 introduced a widespread process of democratization within the Portuguese school system, notably with the establishment of comprehensive schools (*unificação do ensino*) for twelve- to fifteen-year-old pupils; *liceus* and vocational schools as distinct institutions for this age group disappeared. These structural changes encouraged a form of "school explosion" associated with "democratic claims for equality" in the 1970s.[51] Coeducation—reintroduced with the Veiga Educational Act (1972)—was generalized to all levels of schooling. The creation of comprehensive schools opened opportunities, particularly for the lower classes, to pursue higher levels of education. Statistical data for the following decades show that girls continued to outnumber boys, although feminists stressed that girls were integrated into a male educational system where *the male pupil* and *the male teacher* were seen to represent humanity.[52]

The Portuguese entry into the European Union in 1986 coincided with the passage of the *Lei de Bases do Sistema Educativo* (Comprehensive Law of the Educational System), which contributed to the consolidation and diversification of secondary education from the 1980s onward. This law defined the *secondary* level of the educational system as consisting of three non-compulsory years (10th, 11th and 12th grades) while the previous nine years constituted "basic education" (*educação básica*). These fundamental modifications coincided with an expansion in school attendance encouraged in part by the reemergence of vocational schools. For girls, these changes were accompanied by a renewed emphasis on coeducation as the *natural* pathway to implement equality between the sexes—only a few private secondary schools remained single-sex. In quantitative terms, the 1990s witnessed the presence of girls in greater numbers in secondary as well as higher education. Women represented one quarter of all university graduates in 1960, and by the mid 1990s they were 62.9 percent of all graduates. Despite extending compulsory education later than in most EU countries, Portugal now ranks among the leaders in terms of the number of women in higher education, with a proportion of 130 women for 100 men.[53] This achievement has generated, as elsewhere, a certain amount of backlash; notably, political discourse has blamed boys' disaffection from schooling on girls' educational success.

The Extension of Secondary Education to the Former Colonies

It was mainly during the colonial wars in the 1960s, under the pressure for economic development, that the colonial states started to expand high schools and

vocational schools. In colonial Mozambique the first girls' high school was established in the then capital Lourenço Marques, and two mixed high schools were
opened in Quelimane and Nampula in 1961. Previously only three high schools
existed in Mozambique, two of them established in the 1950s.[54] In Angola, the first
boys' high school was established in Luanda in 1919, but mixed-sex high schools
only appeared and expanded in the 1960s. Until now, however, the specific place of
girls in this educational offering had not received attention. Future research must
focus on the relationship between empire, diaspora, the provision of secondary education for girls, and the construction of gender identities in the pursuit of local
occupations and professions.[55]

Conclusion

While girls' secondary education in Portugal has received some scholarly attention
over the decades, much scholarship remains to be done: on women educators and
pedagogues, on their activities opening secondary and other schools, on the development of coeducation, or the gendered historical experiences of colonial education.
Even recent edited studies on women in Africa or colonial education completely
neglect this subject.[56] Clearly, the conservative political context for much of the
twentieth century has affected the pursuit of research on such issues. Only recently
have women's studies and women's history developed in a few research centers.
A fragmented perspective remains on girls' secondary education, and the impression
one gets is that the road to a more complete and systematic perspective will require
long and arduous work. Contemporary agendas should encourage scholars not only
to explore the significance of education for women within their respective societies,
but also challenge historians of education to reconceive traditional narratives paying
greater attention to issues of difference and diversity.

Notes

1. Ana de Castro Osório, *Às Mulheres Portuguesas* (Lisbon: Livraria Viúva Tavares Cardoso, 1905).
2. Áurea Adão, *As Politicas Educativas nos Debates Parlamentares: o caso do ensino liceal* (Porto: Afrontamento, 2001), 496.
3. Vasco Pulido Valente, for example, focuses on the creation of the Portuguese high school (liceu) and emphasizes the curriculum, pupil enrolments and pedagogical discourses. See *O Estado Liberal e o Ensino. Os Liceus Portugueses (1834–1930)* (Lisbon: GIS, 1973). For more recent historical work that includes analyses of both boys' and girls' "liceus," see António Nóvoa and Ana Santa-Clara, eds., *Liceus de Portugal—Histórias, Arquivos, Memórias* (Porto: Asa, 2003); João Barroso, *Os Liceus—organização pedagógica e administração (1836–1960)* (Lisbon: Gulbenkian/JNICT, 1995).
4. Irene Vaquinhas, "O Real Colégio Ursulino das Chagas de Coimbra. Notas para a sua História," *Revista Portuguesa de História* 31, 2 (1996): 427–447; and "Alguns aspectos da vida quotidiana num colégio feminino no século XIX: o caso do Real Colégio Ursulino das Chagas de Coimbra (1874–1880)," *Revista Gestão e Desenvolvimento* 5–6 (1997): 213–247.

See Zulmira Santos on convent education, "Para a historia da educação feminina em Portugal no século XVIII: a fundação e os programas pedagógicos das visitandinas," in *Estudos em Homenagem a Luís António de Oliveira Ramos*, Universidade do Porto, Faculdade de Letras (FLUP/UP) (2004): 985–1001; Aurea Adão, *Estado Absoluto e Ensino das Primeiras Letras—as escolas régias (1772–1792)* (Lisbon: F. Calouste Gulbenkian, 1997); Rogério Fernandes, *Questionar a Sociedade, interrogar a História, (re)pensar a Educação* (Porto: Afrontamento, 2004); Maria Antónia Lopes, *Mulheres, Espaço e Sociabilidade. A transformação dos papeis femininos em Portugal (segunda metade do século XVIII)* (Lisbon: Livros Horizonte 1989).

5. Graça Abranches, "Homens, mulheres e mestras inglesas," in *Entre Ser e Estar—raízes, percursos e discursos da identidade*, eds. M. Irene Ramalho and A. Sousa Ribeiro (Porto: Afrontamento, 2002), 255–305; Helena C. Araújo *Pioneiras na Educação—as Professoras Primárias na Viragem do Século: Contextos, Percursos e Experiências, 1870–1933* (Lisbon: Instituto de Inovação Educacional, 2000); and "Pathways and Subjectivities of Portuguese Women Teachers through their Life Histories, 1919–1933," in *Telling Women's Lives*, ed. Kathleen Weiler and Sue Middleton (Middlesex: Open University, 1999), 113–129.

6. Sérgio Grácio, *Política educativa como tecnologia social—as reformas do ensino técnico de 1948 e 1983* (Lisbon: Horizonte, 1986), 15.

7. Teresa Pinto, *O ensino industrial feminino oitocentista—a escola Damião de Góis em Alenquer* (Lisbon: Colibri, 2000).

8. Cristina Rocha, "A Educação Feminina. Entre o Particular e o Público, o Ensino Secundário Liceal nos anos 30," MA thesis, Universidade Nova de Lisboa, Faculdade de Ciências Sociais e Humanas, 1989.

9. Irene Pimentel, *História das Organizações Femininas no Estado Novo. O Estado Novo e as Mulheres* (Lisbon: Círculo de Leitores, 2000).

10. Maria Manuel Vieira, "Letras, artes e boas maneiras. A educação feminina das classes dominantes," *Análise Social* 120, 28 (1993): 7–53; see Maria José Magalhães *Movimento Feminista e Educação* (Lisbon: Celta, 1998) on the connection between the feminist movement and education.

11. For the work on periodization, see Laura Fonseca, "Contornos da escolarização das raparigas em Portugal: olhar sócio-histórico para resignificar as mudanças educacionais e uma nova agenda de género em educação," in e*x-aequo* 15 (2007): 69–87.

12. Antonio Nóvoa, *Le temps des professeurs* (Lisbon: INIC, 1987), 117.

13. Pupils applied separately to the individual courses of the *aulas régias* that were not organized as a system.

14. See Rogério Fernandes, "Génese e consolidação do sistema educativo nacional" in *O Sistema de Ensino em Portugal (sécs. XIX–XX)*, ed. M. Candida Proença (Lisbon: Edições Colibri, 1998), 23–46.

15. Rogério Fernandes, *O Pensamento Pedagógico em Portugal* (Lisbon: ICT, M.E.C, 1978), 41.

16. Luís António Verney was an Oratorian. He died in exile in Rome. His main work was published in secrecy, since the Inquisition forbade it.

17. Ribeiro Sanches was of Jewish origin and studied at Coimbra and Salamanca. He lived in exile in England, France, and Russia where he was a physicist at the court.

18. Rómulo de Carvalho, *História do Ensino em Portugal- desde a fundação da nacionalidadeaté ao fim do regime Salazar-Caetano* (Lisbon: F. Calouste Gulbenkian, 1986), 417.

19. Fernandes, *O Pensamento Pedagógico*, 70.

20. Sanches, *Cartas sobre a Educação da Mocidade* (Porto: Ed. Domingos Barreira, n.d.), 190–191; also Teresa Joaquim, *Menina e Moça—a construção social da feminilidade* (Lisbon: Fim de Século, 1997).

21. In 1622, in Portugal, there were 450 men's and women's convents. In 1763, after the expulsion of the Jesuits, this number had increased to 538, of which 131 were women's. P. Oliveira, *História Eclesiástica de Portugal* (Lisbon: União Gráfica, 1940), 268–269.

22. Santos, "Para a história da educação feminina."

23. Adão, *Estado Absoluto*, 188.

24. Vaquinhas, "O Real Colégio Ursulino," 430.

25. Private collection, unpublished diary, 1899.

26. Elzira Machado Rosa, *Bernardino Machado, Alice Pestana a educação da mulher nos fins do sec. XIX* (Lisbon: Cadernos da Condição Feminina, 1989).
27. Araújo, *Pioneiras*. Novels also refer to well-qualified Portuguese governesses from socially disadvantaged groups. These women were often better accepted than foreign governesses, who were seen as undermining the national culture.
28. Leila M. Algranti, "Conventos e Recolhimentos em Portugal e na América Portuguesa—um estudo comparativo sobre as instituições de reclusão feminina (séculos XVII e XVIII)," in *Desafios da Comparação—Familia, Mulheres e Género em Portugal e no Brasil*, ed. A. Cova, N. Ramos, and T. Joaquim (Lisbon: Celta, 2004), 65–88.
29. This missionary order, founded by Anne-Marie Javouhey, came to Lisbon in 1881 with the intention of opening schools in Angola.
30. Zélia Pereira, "Os Jesuítas em Moçambique. Aspectos da acção missionária portuguesa em contexto colonial (1941–1974)," *Lusotopie* (2000): 81–105.
31. Helena C. Araújo, "The Emergence of a 'New Orthodoxy': Public Debates on Women's Capacities and Education in Portugal (1880–1910)," *Gender and Education* 4, 1/2 (1992): 7–24, and "Mothering and Citizenship. Educational Conflicts in Portugal," in *Challenging Democracy: International Perspectives on Gender and Citizenship*, ed. Madeleine Arnot and Jo-Anne Dillabough (London: Routledge/Falmer Press, 2000), 105–121.
32. Natividade Monteiro, "Maria Veleda (1871–1955)—uma professora feminista, republicana e livre-pensadora," MA thesis, Universidade Aberta, Lisbon, 2004.
33. See, in particular, the encyclicals of Leon XIII in 1891 and that of Pius XI on Christian marriage (*Casti Connubi*) in 1930.
34. Caiel, *O que deve ser a instrução secundária da mulher?* (Lisbon: Typografia Moderna, 1892).
35. M. Borges Graínha, *A Instrução secundária de ambos os sexos no estrangeiro e em Portugal* (Lisbon: Typografia Universal, 1905).
36. Helena Ribeiro de Castro, "Teresa de Saldanha: um projecto pedagógico inovador no século XIX," Ph.D. dissertation, Universidade de Lisboa, Faculdade de Ciências, 2007.
37. Maria Guilhermina Bessa Gonçalves, *A Comunidade Britânica no Porto* (Porto: Afrontamento, 2005).
38. Maria de Lurdes Costa, "Colégio Nossa Senhora da Estrela," MA thesis, Universidade do Porto, Faculdade de Psicologia e de Ciências da Educação, 2007.
39. Nóvoa and Santa-Clara, *Liceus*.
40. James Albisetti found a similar pattern in Spain: "The French *Lycées de jeunes filles* in International Perspective, 1878–1910," in *Paedagogica Historica* 40 (2004), 150.
41. James Albisetti, "Catholics and Coeducation: Rhetoric and Reality in Europe before *Divini Illius Magistri*," *Paedagogica Historica* 35, 3 (1999): 667–696.
42. Valente, *O Estado Liberal*, 78. See also Pedro José da Cunha "O ensino secundário do sexo feminino em Portugal," *Revista de Educação Geral e Técnica* 4, 4 (1916): 224–233.
43. There is not much information on the social origins of girls attending the first *Liceu*. Within the former school Maria Pia, however, references to the women primary student teachers in 1884–1885 as both "ladies and girls" suggest that middle-class and upper-working-class women attended it: Araújo, *Pioneiras*, 106.
44. Araújo, *Pioneiras* 108.
45. Valente, *O Estado Liberal e o Ensino*, 104.
46. See the articles by Luís Grosso Correia and Maria Judite Seabra on these two high schools in Nóvoa and Santa-Clara, *Liceus*, 204.
47. Stephen S. Stoer and Helena C. Araújo "A Contribuição da Educação para a Formação do Estado Novo: continuidades e rupturas, 1926–1933," in *O Estado Novo, das origens ao fim da autarcia, 1926–1959* (Lisbon: Fragmentos, 1987).
48. Pimentel, *História das Organizações Femininas*, 34.
49. Nóvoa and Santa-Clara, *Liceus*.
50. Sérgio Grácio, *Política educativa*; Cristina Rocha, "Entre l'Enseignement Privé et l'Enseignement Public"; Helena C. Araújo, "La Feminisation du Lycée au Portugal: de la situation d'élèves à la situation d'enseignantes" (Florence: European University Institute, Centre de Recherche sur la Culture Européenne, EUI Colloquium Papers, 1990).
51. Stephen R. Stoer, *Educação e Mudança Social* (Porto: Afrontamento 1986).

52. Fonseca, "Contornos da escolarização das raparigas em Portugal."
53. António Barreto, *A Situação Social em Portugal 1960–1999*, Vol. II (Lisbon: Imprensa de Ciências Sociais/ICS 2000). See also Ana Nunes Almeida and Maria Manuel Vieira, *A escola em Portugal* (Lisbon: ICS 2006).
54. Zélia Pereira, "Os Jesuítas em Moçambique."
55. See Joyce Goodman, "Their Market Value Must be Greater—Secondary School Headmistresses and Empire, 1897–1914," in *Gender, Colonialism and Education—the politics of experience*, ed. Joyce Goodman and Jane Martin (London/Portland: Woburn Press, 2002), 175–198.
56. See Inocência Mata, ed., *A Mulher em África—Vozes de uma margem sempre presente* (Lisbon: Colibri, 2007) and António Nóvoa, ed., *Para Uma História da Educação Colonial* (Lisbon: Educa, 1996).

Chapter 7

Champion in Coeducation: The Netherlands

Mineke van Essen and Hilda Amsing

"Recently," wrote the Dutch Minister of Internal Affairs to the governors of the Dutch provinces in 1827, "some complaints have been heard about the lack of qualified girls' schools and about the neglect of their education." Therefore, he asked His Majesty's governors to investigate the state of the art regarding girls' schools in their regions and the training of women teachers. This seems to be the first time in Dutch history that the national government expressed any special concern for the education of its girls. Except for providing ten grants to prepare girls for teaching by attending an excellent secondary girls' boarding school; however, nothing actually changed.

This early concern was reported in the first Dutch dissertation about girls' education, published in 1985.[1] The origins of this study lie in 1979, when the fledgling Dutch association of women history students organized a small conference on the subject of girls' education. The conference papers can be considered the beginning of modern research into this subject. During the nineties, scholarship in the field burgeoned: in addition to twenty-seven articles, six dissertations, three books, a special volume of a respected Dutch journal on education, and an anthology were published, almost all of them in Dutch.[2] The majority of these dealt with general secondary education. In 2000, gender in education and upbringing was the theme of the first yearbook of the Flemish-Dutch association of historians of education. The historiographical article in this yearbook was proudly entitled: "From the Margin to the Center."[3] Since then, however, due to the increasing lack of staff research, the number of publications has drastically decreased.

The chronological focus for all research concerning Dutch (and particularly girls') secondary education lies in the period between two secondary school acts, one in 1863 and the other in 1963. Almost all research has centered on the three highest forms of secondary education, meant for children of the elite, and the middle- and upper-middle-class: the *gymnasium*, the Higher Burgher School (*Hogere Burgerschool*,

HBS), and the secondary girls' school (*Middelbare Meisjesschool*, MMS). Almost no attention has been paid so far to the lower forms of secondary education, notably, higher primary schools (*Meer Uitgebreid Lager Onderwijs*, MULO) meant for lower middle-class pupils. The complicated status of the MULO, combined with the dominance of coeducation in the Netherlands, explains this scholarly neglect; indeed, few gender-specific MULOs existed (except for Catholic ones). Coeducation also explains the scarcity of available documentation about women secondary teachers. No special training courses were set up for them and most women teachers worked in mixed schools. Even less is known about girls' secondary education in the Dutch colonies (the Netherlands East Indies, Surinam, the Dutch Caribbean). Consequently, this chapter concentrates mainly on secondary education for Dutch girls from the upper classes.

Enlightenment Influences—Eighteenth Century

Three famous female authors represent the Dutch Enlightenment with respect to the position of women and their education: Belle van Zuylen (outside the Netherlands better known as Isabelle, or "Madame" de Charrière), Betje Wolff, and Aagje Deken. In their own way each of them rebelled against the traditional female role. Van Zuylen (1740–1805) came from a noble family and was extremely talented. She received a good education that included Latin and the sciences, thanks to being allowed to attend her brothers' lessons given by a private tutor. She refused to be a traditional elite spouse. Instead, she followed her heart in love affairs and ended up playing a role in the European culture of her time. Publishing in French, she became more famous outside the Netherlands than in the country of her birth.[4] Betje Wolff (1738–1804) came from a middle-class family and ran away at seventeen with a boy friend. After returning home, she chose to marry a vicar thirty-one years her senior, whom she respected very much. She then began to write and, through her publications, came into contact with the lower-middle-class Aagje Deken (1741–1804), who had been raised in an orphanage. They became coauthors.

All three protested against the lack of education for girls and women in their day. Inspired by the enlightenment ideas of the French Revolution, Wolff and Deken took part in the late-eighteenth-century Dutch democratic movement to change society, which included changing the position of women. In contrast to a minority of radicals who proclaimed equal rights between the sexes, the majority of the reformers, Wolff and Deken included, advocated a more moderate view on the elevation of women. Referring to nature, they argued that men and women were created as fundamentally different creatures who needed to make their own gender-specific contribution to society. Woman's main responsibility was to educate her sons to become virtuous and responsible Dutch citizens, and her daughters to be good mothers.[5] To fulfill this duty properly, Wolff and Deken called for better education for girls. This should, however, never turn them into *savantes*, or "blue-stockings." Their bourgeois ideas about girls' education differed fundamentally from those cherished by Van Zuylen, who was herself a *savante*.

Ideal and Reality: The First Half of the Nineteenth Century

Partly due to the social classes to which they belonged, Wolff and Deken and Van Zuylen represented two conflicting concepts of girls' education—gender-specific education for motherhood versus gender-neutral intellectual education—which continued to be seen throughout the centuries that followed. However, early nineteenth-century reformers did start off with one common goal, namely, the improvement and extension of formal schooling for girls.

Traditionally, girls as well as boys from all classes, except the elite, irregularly and voluntarily attended coeducational elementary schools in order to learn to read and to be instructed in (Protestant) religious matters. Often the teachers were not well-equipped for their task, if only because they had also other church-related functions, including as sextons. Bourgeois children (though still far more boys than girls) also learned writing and arithmetic. Sometimes they went on to a single-sex "French" school where they received an extended education including French, at that time the *lingua franca*. French schools for girls mainly taught French and embroidery. Latin schools (the *gymnasium*) in the cities were intended for upper-class boys and prepared them for university. Elite children received their primary education at home from a governess (in most cases, French or Swiss). For boys, male tutors were also hired to teach specialized subjects such as classical languages and mathematics. Some girls, who were privileged like Van Zuylen, were permitted to follow those lessons. To complete their education, elite girls were sometimes sent to a French boarding school to learn good manners and accomplishments, dancing, embroidery, music, and whatever else was considered necessary for elite spouses.

One of the most impressive results of the enlightenment reformers was the introduction of the 1806 National Primary School Law. Although attending school had not yet become compulsory, the law did introduce public primary schools for all children, girls and boys alike, whose parents could not afford to send them to private schools or to provide a home education. The first step on the route to becoming better-educated women was thus taken. This was not sufficient, however, to make up for the lag in girls' education, hence the complaints from 1827 mentioned in the introduction. Until the 1860s, the education of girls remained a low priority in preindustrial Netherlands. Attending a secondary school was a privilege for upper and—sometimes—middle-class children, although even in those social classes this was exceptional for girls. In addition, there were only a handful of women teachers. In 1850, of the approximately 4,000 public elementary and secondary school teachers, 2 percent were women. In private schools, 27 percent of the 1,621 school teachers were women. The majority were members of Catholic sisters' congregations teaching at girls' schools.[6]

This uneven gender rate resulted from the dominance of coeducation in primary schools. Only Catholics (at that time about 30 percent of the Dutch population) and parents from the elite preferred their daughters to be educated at girls' primary schools. At that time women could only teach in girls' schools, because they were considered unsuitable to teach boys. In the domain of girls' schools, however, women had the same opportunities as men. Almost all public and private girls' schools had a woman at the helm. Gender barriers relative to promotion or income did not exist. Neither does there appear to have been resistance to married women teachers. In the 1830s, a quarter of all women teachers were either married or widowed. Among them was the most famous woman of the first half of the nineteenth century (outside the royalty), Anna Barbara Van Meerten-Schilperoort (1778–1853). By heading a prestigious and expensive private day- and boarding school and publishing a number of books, she dominated the gender discourse on education in her day. In her educational novels for women and adolescent girls, she shows herself an adherent of Wolff and Deken's concept for girls' education. She taught her audience to become devoted mothers and wives, who were well educated but who never aspired to be *savantes* or to play a role in public life.[7]

The "Burning Question": The 1870s National Debate

In the 1860s, complaints about the poor quality of girls' secondary education intensified as a result of three developments. First, the Netherlands witnessed a suddenly increased concern about the number of young women from the middle classes who did not marry. Although this observation might have been unfounded, the belief required a response. Single bourgeois women began to feel pressure to find paid work. The traditional French girls' schools, however, being their sole educational opportunity, did not offer suitable preparation. Second, feminist ideas regarding women's right to participate in the public domain began to emerge slowly. This development was stimulated by international discussions, for example, by John Stuart Mill's book on the subjection of women, which was translated into Dutch in 1870. The improvement of women's education was judged to be of prime importance in the realization of emancipation. However, the decisive factor arose from one educational development, namely, the first Dutch Secondary School Act in 1863. This act instituted a new public secondary school for boys, the Higher Burgher School or HBS, offering a modern curriculum of sciences and modern foreign languages. Its standard was guaranteed by a state-controlled examination. For girls, the law mentioned the possibility of founding private secondary schools for girls offering a more in-depth education than the traditional French girls' schools. The law, however, neither formulated a curriculum nor introduced a state examination. By adding to the educational opportunities of boys, while almost neglecting the secondary education of girls, the law made gender differences more manifest.

These circumstances provided the impetus for a national debate on what suddenly came to be called *la question brûlante*, the burning question of whether girls should have a decent secondary education, and, if so, what should be its aims and content. The most radical position in the debate was taken by an obscure woman writing under the pseudonym Marie Delsey. She promoted the idea that girls should be admitted to the same secondary schools as boys, "in order to develop reason and sensibility together with their brothers and be educated to fulfill the task that awaits them in society." Most participants in the debate followed the more moderate position of the prominent male secondary school inspector D.J. Steyn Parvé: girls should receive the same general secondary education as boys, but not to the same depth. Especially when it concerned mathematics, sciences, and economics, the basics should suffice. His argument was that, apart from a few exceptions, all women would become wives and mothers. Some women teachers took part in the debate on the margins but none in a leading position. Instead, Steyn Parvé's curriculum design became the model for all twelve private secondary girls' schools established on the basis of the 1863 act (MMSes) that existed by the end of the nineteenth century. In addition to Dutch and other modern languages (French, German, English), the curriculum consisted of history, geography, and arts and crafts subjects such as music, drawing and, of course, needlework. Mathematics and sciences were limited to a primary-level introduction.[8]

However, Delsey's position in this debate did not go unheard. As a result of an official request in 1871 from the physician Jacobs to the Minister of Internal Affairs, his daughter Frederika (a younger sister of Aletta Jacobs, the first Dutch medical doctor and the Netherlands' most famous feminist) was admitted to an HBS. In that same year nine other girls followed her example. Initially, admission was meant to be the exception in cases where there was no available girls' school. Despite the fact that until 1907 each individual girl needed to obtain permission from the authorities, attending a boys' HBS or gymnasium became a popular option. At the end of the nineteenth century one-third of all bourgeois girls who received a secondary education attended a coeducational school. After 1900 they steadily outnumbered students at girls' schools; in 1925 the percentage had increased to seventy-four.[9]

Relatively speaking, the number of girls enrolled in secondary education was low. In 1900 school attendance was only compulsory up until the age of twelve, with the result that only about 4 percent of all children aged twelve to nineteen received some form of general secondary education during this period; and only about 0.1 percent of the girls in this age group went to a coeducational school. In contrast to MMS girls, until the beginning of the twentieth century these girls were taught exclusively by men.

Revival of Gender Specificity, 1920–1965

After Dutch women won the vote in 1918, the feminist movement lost its fighting spirit. Moreover, due to modern scientific developments in psychology, particularly those brought forward by Darwinist psychologists such as Stanley Hall, the majority of feminists advocated differentiation instead of equality. Modern civilization, they believed, encouraged the development of typically feminine occupations and also a reevaluation of motherhood as the true destiny of women. Consequently, a modern girls' education should prepare girls for their specific vocation. It was not to be coeducational but gender-specific.

Despite the dominant objections to coeducation at the time, the participation of girls in coeducational schools continued to increase, both in secondary as well

Table 7.1 Numbers and percentages of girls in coeducational secondary schools, per category, 1885–1965

	HBS		Gymnasium		MULO (Higher Primary Schools)	
	Number	%	Number	%	Number	%
1885	80	3	0	0		
1905	1305	17	421	14		
1925	5362	25	1762	43		
1945	8639	22	3267	31	55074	47
1965	14176	28	7072	35	140533	51

Source: CBS, Statistics Netherlands.

as in higher primary education. Table 7.1 depicts the percentages of girls in the classrooms of the different schools. In addition, women teachers had begun to enter these schools. Financial and political circumstances were partly responsible for the continuing popularity of coeducation. The economic problems caused by the First World War and its aftermath, the economic crisis of the 1930s, followed by the Second World War, and the subsequent restoration period frustrated all forms of educational change. In addition, attending an MMS was more expensive because this kind of school received no national state subsidy.

Yet, some signs indicate that the changing ideology did have an effect after all. First, the MMS curriculum underwent alterations in a more feminine direction. Between 1873 and 1947 the percentage of hours devoted to mathematics and sciences decreased from twenty-eight to fifteen, while hours for arts and crafts increased from eighteen to twenty-four. Second, some MMSes added an all-girls HBS or gymnasium department to their institutions; in this department extra time was spent on "subjects particularly important for girls, such as needlework, art history, and hygiene." An additional advantage was that these departments were eligible for state subsidy while MMSes were not. Especially, this construction became an interesting option for the Catholic segment of the population who considered coeducation to be a danger to morality. First, from the 1920s the Dutch State began to subsidize private secondary schools and all-girls' HBS and gymnasium departments at the same level as public schools under certain circumstances. Second, the 1929 papal encyclical *Divini Illius Magistri* condemned coeducation, so Catholic parents who wanted a secondary education for their daughters had no choice but to send their girls to all-girls' institutions.

Until the end of the Second World War, however, the number of religious secondary girls' schools remained low. After the MMS finally received national state subsidy in 1948, however, there was a large increase in both the number of schools and of pupils. The number of MMSes grew from 38 in 1945 to 186 in 1965; and the number of pupils increased nearly tenfold. Eighty percent of these pupils attended private, often Catholic, schools. This increase was caused in part by the rise in age of compulsory education from twelve to fourteen; and because relatively more girls also attended gender-specific secondary education. Most lower-class girls, however, attended a vocational school.

Twentieth-Century Colonial Education of Indigenous Girls

By 1900 the "colonial question" had become a hotly debated political issue. One of the themes discussed was the importance of girls' education as a means of reeducating colonial society. The only so-called native woman involved in this debate was Raden Ajeng Kartini (1879–1904), a young woman from an aristocratic Javanese background, who would become the icon of Indonesian feminism. Until her tragic death in childbirth, she made the case for progress through girls' education in her many letters: "Give Java fine, intelligent mothers and the improvement, the raising of a people, will be but a matter of time."[10]

Kartini's own education began with a European governess, followed by a few years at a European elementary school which she left at age twelve. In the ensuing years she seemed to have had a conventional Javanese and Muslim education at home, although she kept up her acquaintance with Western education through her brother's textbooks. She then met the wife of a Dutch colonial official who introduced her to contemporary Western social and feminist debates. In the years that followed, she devised plans to free herself from the traditional obligations of women by, among other things, pursuing further education in the Netherlands as her brother had done. But both colonial and Javanese obstructions placed on her personal aspirations proved ultimately too strong and, in the end, she could not escape an arranged marriage. After her death, Dutch proponents of colonial reform founded a total of seven modern elementary "Dutch" schools for upper-class Javanese girls called *Kartinischools* in honor of her contribution to the emancipation of Indonesian women.[11]

At the time, the education of Kartini and her brother was exceptional for Indonesian children. Dutch colonial education in the Netherlands East Indies was organized along dualistic lines, with separate schools for Europeans (including some limited admission for Chinese inhabitants) and for natives. Native schools offered a basic curriculum and teaching took place in the local language instead of in Dutch. Most native children did not go to school at all. During the first decades of the twentieth century, education expanded as part of the new Dutch policy to improve the welfare of the indigenous peoples. An intricate web of primary schools evolved: both private and public schools, along with native schools, and Dutch schools meant for Europeans, for Chinese, and for natives, respectively. Different secondary schools for general and for vocational education were also established, although usually pupils from the three population groups remained separate. This expansion extended to girls. Between 1920 and 1930 the literacy rate among Indonesian women increased more rapidly than that of men. On Java, it went from 9 to 13 percent. This was an urban phenomenon, however, mainly confined to the elite who preferred to give their children a European education and to participate in Dutch culture, and to use

Table 7.2   Students in secondary and higher primary education, Dutch East Indies, 1931

Schools	Number of Schools	European (Dutch)		Indonesian		Chinese	
		Male	Female	Male	Female	Male	Female
HBS (public)	6	1682	847	148	73	227	116
AMS (public)	7	127	8	712	52	194	26
Public MULO	35	841	802	4383	1191	739	286
Private MULO	30	950	963	1446	308	436	209

Source: J.E.A.M. Lelyveld, "...Waarlijk geen overdaad, docht een dringende eisch...Koloniaal onderwijs en onder-wijsbeleid in Nederlands-Indië 1893–1942," Ph.D.diss. University of Utrecht, 1992, 191–193.

Dutch as their daily language. Indonesian girls also enrolled in secondary education (see table 7.2), but given the population of 60 million Indonesians (versus 240,000 Europeans), their numbers were negligible. In addition to the HBS and the MULO, a General Secondary School (AMS) was introduced, which was created especially for Indonesian pupils, with a curriculum that attempted to build a bridge between European and Indonesian culture.

One of the Indonesian girls who managed to make use of the scarce educational possibilities was Soewarsih Djojopoespito (1912–1977), who began her education at a Javanese Kartinischool. After graduating from this school, she first went to a MULO, then, with a scholarship from the Kartinifund, to the Dutch primary-teacher-training school in Soerabaja. Conscious of her exceptional position, she became increasingly nationalistic. Private nonsubsidized Indonesian elementary schools with a Dutch curriculum were founded during this time, as a national-istic protest against the Dutch educational system that impeded the entrance of Indonesians into the dominant Western culture. She taught at one of these schools in the 1930s, together with her husband Soewarshi.[12]

In contrast to Dutch children in Indonesia, who were sent to the Netherlands for university or higher vocational education after secondary education (sometimes even earlier), only a handful of Indonesian children had a chance of doing so. These mostly went to the Netherlands for university, higher technical or teacher training. According to a rough estimate, only 150 Indonesians were studying in the Netherlands in 1930, of whom only a small minority was female. Had the country not become independent in 1949, these numbers would probably have increased after the Second World War, as happened with indigenous girls and boys in the Dutch colonies in the Americas. Most of them had also completed a secondary education at home and only came to the "motherland" for further study. These small territories—in 1930 Surinam had a population of about 150,000 inhabitants, the Dutch Antilles, far less—offered fewer possibilities for secondary education than did the East Indies. Until the Second World War, only a few MULOs were established; later, Surinam was also provided with an AMS. Figures for the gender ratio in these schools (which, except for the Catholic ones, were all coeducational) are not available.

The "Burning Question" Revisited: The 1970s Feminist Movement and Its Aftermath

Against the backdrop of the increasing postwar popularity of the MMS, it is surprising that the new secondary school act of 1963, designed by a Catholic Minister of Education, abolished gender differentiation. A broad curriculum for all, irrespective of gender, now became the sole standard for secondary education. Consequently, all MMSes disappeared. By the 1970s, girls had come close to equaling boys in educational participation, although actual equality had not yet been reached. Since the act allowed for a great deal of freedom in the choice of subjects, individual choices still reflected gender difference. Far more girls than boys chose a so-called "fun package": a curriculum consisting entirely of languages, history, and arts and crafts, thus avoiding mathematics and sciences.

The new feminist movement of the 1970s criticized this common practice, leading to a revival of the debate about the "burning question." The previously assumed gender neutrality of coeducational schools was also questioned. Scholars and feminists denounced a "hidden curriculum" around gender. In particular, they showed that teachers paid less attention to girls than to boys and acted according to traditional stereotypes, privately assuming that girls were less intelligent than boys or would choose "feminine" occupations. Textbooks also expressed stereotypical ideas, while history books neglected the role of women. In addition, campaigns were started to motivate girls to increase their level of education and to choose not just the traditional female subjects and occupations. These combined activities produced a miraculous result. Together with the rise in the age of compulsory education to sixteen in 1975, this brought about a further increase in the enrollment of girls in secondary education (as well as in higher vocational and university education). Two other, more contemporary, results are noteworthy. First, in 1990, and later in 1991, women's history became an obligatory subject for examination for all pupils who opted for history. Three years later, the subjects Care and Technology were introduced into the compulsory basic secondary school curriculum.[13]

In addition to the traditional middle-class target group for general secondary education, girls from the lower classes also started attending these schools. Consequently, class-related inequality in education decreased. With the large-scale arrival of immigrants to the Netherlands, particularly those from Morocco and Turkey, however, new differences replaced old ones. The participation in secondary education by children from these groups lags far behind, although, in most cases, immigrant girls are doing better than boys, a phenomenon that they share with girls from a native-Dutch background. In 2005–2006, of all of the students in the highest grades of general secondary education, 53 percent were female, both immigrant and native-Dutch alike. In addition, the differences in subject choices between boys and girls are decreasing, although girls still opt less frequently for mathematics and sciences.[14] As a result, only the rudiments of the two eighteenth-century concepts of girls' education can still be discerned.

Notes

1. Mineke van Essen, *Onderwijzeressen in niemandsland* (Nijkerk: Intro, 1985).
2. Among these publications a survey of Dutch girls' education from 1800 to the present which has still not been superseded: Mineke van Essen, *Opvoeden met een dubbel doel. Twee eeuwen meisjesonderwijs in Nederland* (Amsterdam: SUA, 1990). Most of the information in this chapter is based on this book.
3. Mineke van Essen, "Van marge naar middelpunt. Twee decennia Nederlands historisch onderzoek naar sekse in opvoeding en onderwijs," in *Genderconcepties en pedagogische praktijken. Jaarboek voor de geschiedenis van opvoeding en onderwijs 2000*, ed. van Essen (Assen: Koninklijke Van Gorcum BV, 2000), 6–40.
4. C.P. Courtney, *Isabelle de Charrière (Belle de Zuylen): A Biography* (Oxford: Voltaire Foundation, 1993).
5. W.Ph. te Brake, R.M. Dekker and L.C. van de Pol, "Women and Political Culture in the Dutch Revolution," in *Women and Politics in the Age of Democratic Revolution*, ed. Harriet B. Applewhite and Darline G. Levy (Ann Arbor: The University of Michigan Press, 1990), 109–147.
6. Mineke van Essen, "Strategies of Women Teachers 1860–1920. Feminization in Dutch Elementary and Secondary Schools from a Comparative Perspective." *History of Education* 28, 4 (1999): 413–433.
7. Mineke van Essen, "Anna Barbara Van Meerten-Schilperoort (1778–1853). Feminist Pioneer?" *Revue Belge de Philologie et d'Histoire* 77, 2 (1999): 383–401.
8. Mineke van Essen, "'New' Girls and Traditional Womanhood. Girlhood and Education in the Netherlands in the Nineteenth and Twentieth Century," *Paedagogica Historica* 29, 1 (1993): 125–151.
9. Nelleke Bakker and Mineke van Essen, "No Matter of Principle. The Unproblematic Character of Coeducation in Girls' Secondary Schooling in the Netherlands, ca. 1870–1930," *History of Education Quarterly* 39, 4 (1999): 454–475.
10. Joost Coté, "Celebrating Women's Labour. Raden Ajeng Kartini and the Dutch Women's Exhibition, 1898," in *Een Vaderland voor Vrouwen*, ed. Maria Grever and Fia Dieteren (Amsterdam: Stichting beheer IISG / VVG, 2000) 119–135, quotation page 131.
11. Frances Gouda, *Dutch Culture Overseas: Colonial Practice in the Netherlands Indies, 1900–1942* (Amsterdam: Amsterdam University Press, 1995), 75–117.
12. G.P.A. Termorshuizen, "A Life Free from Trammels: Soewarsih Djojopoespito and Her Novel 'Buiten het gareel,'" *Canadian Journal of Netherlands Studies* 12, 2 (1991): 30–37.
13. Geert Ten Dam and Monique Volman, "Care for Citizenship. An Analysis of the Debate on the Subject Care," *Curriculum Inquiry* 28, 2 (1998): 231–247.
14. Annemarie van Langen and Lyset Rekers-Mombarg, "Group-related Differences in the Choice of Mathematics and Science Subjects," *Educational Research and Evaluation* 12, 1 (2006): 27–51.

Chapter 8

Politics and Anticlericalism: Belgium*
Eliane Gubin

In Belgium, as elsewhere, girls' education constituted a critical step toward autonomy. Struggles to gain access to secondary education acquired a particular intensity,
however, given the political context and the priority Belgian feminism accorded to
women's economic emancipation.

The relationship between education and politics has traditionally dominated historical research due to the bitter struggles between Catholics and liberals (and then
socialists) over school control, notably in two "school wars" (1878–1884 and 1954–
1958). In addition to these ideological, pedagogical, and financial battles, language
issues further complicated debates about schooling once the clerical-secular struggles had waned.[1] As a result, the bibliography on the question is enormous and
includes studies on the organization of schooling in both Belgium and the colonies,
monographs about individual institutions, biographies of pedagogues, analyses of
teaching personnel, but very few studies that include a gender dimension.[2] Even
fewer studies address girls' secondary education, which only affected the middle
and upper classes in the nineteenth century. Recapturing the place of women in the
educational transformations of the past 150 years involves expanding the focus to
include the progressive democratization of society and its socioeconomic evolution,
especially since young women arrived massively in the workplace in the twentieth
century. As a result, a gender perspective of these phenomena involves an attention
to political, economic, and social history, as well as the history of pedagogy, women,
and feminism.

To date, most research on girls' secondary education has centered on pioneering women and their schools within general education, where the stakes were more
spectacular, since these studies allowed women to break into the university. Other
pathways to education, such as the normal schools for future schoolteachers and
régentes,[3] or technical and professional training, have been less studied, despite the
fact that gendered differences in programs and objectives were very apparent and
persist to this day.

This chapter examines three quite unequal time periods: the "long nineteenth century," which saw the emergence of a model of girls' secondary education; the interwar period, when this model became more visible as well as more diverse; and the second half of the twentieth century until Belgium was converted into a Federal State (between 1972 and 1992), which represents, as elsewhere, the period when secondary education acquired a mass student body.

The "Long Nineteenth Century"

The revolution, which abruptly ended Belgium's union with the Netherlands (1814–1830), ushered in a new constitution in 1831 that guaranteed the freedom of both education and association. The nonintervention of the State[4] constituted a founding principle that immediately allowed the Catholic clergy to reestablish a vast network of religious schools lost during the Dutch regime.[5] This development of Catholic education continued throughout the nineteenth century due to the pressure of the school wars, which intensified at mid-century when the State sought to reestablish control over education. New tensions emerged at the beginning of the twentieth century, thanks to the sudden appearance of French religious congregations seeking new clientele following the anticlerical laws that banned them from teaching in France (Combes laws, 1901–1904).

Girls' Education, a Political Struggle?

The existence of a dense network of private Catholic schools, many of which for historic reasons received public financial support, weighed heavily on girls' education. During the first half of the century, women religious held a virtual monopoly over the secondary education of young bourgeois women, whom they raised to become good Christians, wives, and mothers.

Tension between forms of modernity favorable to women's emancipation and conservative forces defending traditional gender roles affected the evolution of girls' education. The few existing feminists at mid-century were forced to seek alliances within anticlerical and Masonic circles after 1846 when secondary education became a political stake in the struggles between liberals and Catholics. For liberalism in general, the effort to wrest women from the hold of religion constituted an essential component in the struggle to weaken the power of the Church. Girls' emancipation was not, however, the object of this struggle: most liberals paid short shrift to the idea of individual self-development and even less to that of the equality between the sexes. Their goal was to rid girls' education of its central focus on the catechism and to create schools that would prepare women to defend their husbands' (liberal) ideas. Their vision of women's social role remained traditional: as domestic helpmates women played a critical role in the family, showcasing their husbands' success and ensuring the family's biological and ideological continuity.

Feminists, including Léonie de Waha (1836–1926) and Isabelle Gatti de Gamond (1839–1905), needed the support of liberal politicians, however, in order to create a secular system of secondary education for girls, similar to what existed for boys. They advocated a system that would disrupt traditional gender roles in bourgeois families, in addition to challenging contemporary stereotypes concerning women's inferior intellectual capacities, nourished both by the Church and positivist sciences at mid-century. Feminists organized their struggle along two lines: first, the creation of secondary institutions for girls; and second, the defense of more rigorous studies similar to those in *athénées*.[6] Ideally, feminists argued that this development should be accompanied by coeducation—a position also supported by a few enlightened pedagogues—but this latter question rubbed too much against the conviction that it was necessary to separate the sexes in order to preserve morality.

The convergence between the views of the feminist movement and those of the liberal party resulted in a number of strategic alliances beginning in the 1860s, which allowed the emergence of a new secular form of girls' secondary education.

Until 1864: Educational Resources Beyond State Control

Before the emergence of this more modern and secular version of girls' education, many private boarding schools existed alongside those run by religious congregations.[7] Their quality varied greatly, but both focused on providing the sort of religious education considered indispensable for the acquisition of feminine virtues, modesty and submission. Some of these schools had a good reputation—the Heger institution in Brussels, the Passage Institut in Mons, or the *Maison d'éducation française pour demoiselles* in Liège (the French school for demoiselles)—but in general the absence of any control rendered difficult any judgment concerning the quality of their teaching. Boarding-school advertisements all emphasized the serious nature of their program in terms of reading, mathematics, grammar, literature, and geography; they all highlighted the presence of professors hired to teach foreign languages, sewing and embroidery, and even deportment.[8] These institutions trained young women of the bourgeoisie to become perfect housewives, capable of managing servants, entertaining, and shedding honor on their husbands. Alongside the inculcation of moral qualities, these schools also imparted the modicum of culture necessary for worldly society: the ability to drum out a few pieces on the piano, to draw flowers, to converse in a salon. Such accomplishments took up a great deal of time in the curriculum, to the detriment of any serious scientific knowledge. For most bourgeois families, this boarding-school education was an investment in the matrimonial market, rather than a personal advantage.

A small number of young women, notably from the aristocracy, also received an education at home with governesses or tutors. This experience, when coupled with personal curiosity and an impetus for self-learning, produced a number of quite remarkably cultured women, such as Justine Guillery (1789–1864) or Zoé de Gamond (1806–1854).[9] But these women constituted an exception; most young bourgeois women acquired only superficial learning with no perspective of professional training.

Until the 1850s, the leading classes, no matter what their political leanings, deemed these educational opportunities sufficient. The (rare) criticisms about girls' education came from a small group of women with utopian socialist leanings (Saint-Simonians and then Fourierists). They formed a group around Zoé de Gamond who placed girls' instruction at the heart of her project for social renovation: "The female educational system constitutes the main source of trouble in contemporary society; it renders women's situation even more difficult."[10]

Nonetheless, the impetus for change came primarily from political struggles rather than from this critique of the mediocrity of women's studies or the rejection of sexist stereotypes. Liberals became more fully cognizant of the dangers a profoundly religious education posed for the future educators of their children during the pontificate of Pius IX (1846–1878), whose condemnation of all forms of modern liberties spurred an active opposition. In Belgium the liberal party was pushed by its anticlerical wing, under the influence of the pedagogical ideas of the Ligue de l'Enseignement (1864), to support the creation of a secular model of girls' education, structured around the study of science and the reign of reason. This sudden change in liberal attitudes toward girls' education opened the way for a number of determined pedagogues, such as Isabelle Gatti de Gamond or Léonie de Waha, to find the necessary support within liberal municipal administrations to create high-quality institutions for girls.

The Founding Moment: Isabelle Gatti de Gamond's *Cours d'éducation*

In October 1864 the city of Brussels opened its first *Cours d'éducation pour jeunes filles* (Educational courses for young women), under the direction of Isabelle Gatti de Gamond. While only twenty-five years old, she had inherited the ideas of her socialist mother, Zoé de Gamond, and had already developed a vision of a girls' educational system, which she described between 1862 and 1864 in the pages of her journal, *L'Éducation de la Femme*. These ideas shared much with those of the utopian socialist French pedagogical circles she frequented (the Garcin couple, Marie Pape-Carpantier, Jules Delbrück, Mlle Marcheff-Girard, Elisa Lemonnier) and borrowed from common theories of the time (Fröbel, Pestalozzi), as well as foreign experiences, which she then adapted to the Belgian context. She was particularly enthusiastic about the day courses introduced in Paris by David Lévi-Alvarès.[11]

The *Cours d'éducation* offered for the first time a secondary section for girls modeled on that of boys, without religious courses. Despite the opposition she encountered, Isabelle Gatti imposed this programme without eliminating the so-called feminine branches. She was well aware that most of her students intended to marry and become accomplished mothers and wives, running exemplary households; but she also wished them to be cultivated, educated, and liberal in their political inclinations. Alongside the traditional domestic economy, she added a solid program in French, history, geography, foreign languages, gymnastics, and hygiene. The final three years of study out of six progressively included subjects previously considered masculine, such as Greek and Latin, chemistry, higher mathematics, and the philosophy of history.

From the outset Isabelle Gatti sought to emancipate girls intellectually, a radical position that led to many difficult moments with the municipal authorities in Brussels. She disputed the idea that girls were less intelligent than boys and considered this prejudice a means of maintaining masculine domination: "Woman, just like man, is gifted with an intelligence open to culture and development," she wrote as early as 1862.[12] Later she noted that "the obstinate search to prove the inferiority of one sex compared to the other is inspired by secret thoughts of oppression and exploitation."[13]

Without question, the *Cours d'éducation* were a success whose example was then imitated: in a few years, between 1864 and 1870, approximately twenty similar schools appeared in cities with liberal majorities. Thanks to this success, Isabelle Gatti gradually extended her initial project and multiplied the sections within her institution; she created a normal course for primary teachers in 1876, and one for *régentes* (lower secondary teachers) in 1880. By 1892, she had opened preparatory courses that provided the lessons in the classical humanities necessary to enroll in the university (ruling of 1890). Girls then had to pass an exam before a central jury to prove they had learned the same program as that in boys' *athénées*. The city of Brussels also encouraged the development of other *Cours d'éducation*. In 1876 Henriette Dachsbeck opened a second such institution and in 1908 Lily Carter opened a third; both followed the programs established by Gatti and opened as well a preparatory section for the university (in 1896 and 1913).

Catholics initially responded to these developments through a virulent press campaign and parliamentary obstruction that vainly attempted to discredit these institutions. They then responded by reinforcing the programs offered by women religious, in order to ensure secular teachers did not secure a monopoly in serious female schooling. As a result, certain teaching congregations considerably improved their offering. At the same time, various private initiatives also emerged, such as the school founded in Anvers (1892) by Marie Elizabeth Belpaire, a philanthropist. The competition and emulation that resulted from this multiplication of secondary schooling institutions helped raise the general level, although significant differences continued between lay and Catholic schools; in particular Catholic schools never advocated the equality of education between boys and girls.

Training Teachers

The question of programs constituted only one aspect of the issue; competent teachers were required as well. Until 1874,[14] the training of future teachers was exclusively the domain of sixteen Catholic schools subsidized by the State, whose main concern was to inculcate their students with traditional religious values. Students were required to board, were subjected to rigorous discipline, and had to obey a rigid internal rule-book, which resembled that within monastic orders. The implicit model for this training was that of the woman religious, completely devoted to her "vocation." Intellectual and pedagogical training was rudimentary and student morality was, in general, considered more important than intellectual capacities. This explains why feminists always coupled pedagogical training with their concern to improve the level of girls' studies. In Brussels, Isabelle Gatti herself trained her future teachers with an iron hand, forcing them to pass internal exams and to obtain a teaching diploma beginning in 1869. For her university preparatory courses, she hired professors from the Université libre de Bruxelles (the Free University of Brussels).[15] By the end of the century, she had created a corps of qualified teachers: several women who taught for her had obtained university degrees.[16]

Professional Education

The reform measures affecting girls' secondary education also influenced professional education, in part due to the achievements of the former Saint-Simonian Elisa Lemonnier in Paris (1862). Following her example, the first professional school for girls opened in Brussels in 1865, under the initiative of the banker Bischoffsheim and supported by the same social groups who were favorable to Isabelle Gatti. The Bischoffsheim Institute proclaimed a number of revolutionary objectives for the period: aside from the *secular* nature of its teachings, it declared the importance of a general *theoretical* level of instruction for future women artisans, organized apprenticeship training *outside* of the workshop, and *legitimated* the principle of paid work for women. These objectives constituted a total break from the dominant ideal of the angel of the house. From the outset, the Institute set in place a commercial section to prepare students for accounting and office work (both exclusively masculine jobs at the time) and eschewed any training in domestic economy. Similar schools emerged in Anvers and Mons (1874), in Brussels (1873 and 1876), in Verviers (1886), and so on. By 1900 some forty-eight such institutions existed, of varying quality.[17] Just as in secondary education, this success forced women religious to create similar institutions beginning in 1884; their goal, however, was to allow future wives to supplement the familial economy through work at home, while also teaching them to direct their households "with order, intelligence and economy," rather than training them for work outside the family.

1881: The State Intervenes

The liberals returned to power in 1878 after eight years in opposition and immediately set to work codifying secondary-level education for girls.[18] In 1879 the State created a teaching degree for *régentes* and developed normal training for primary schoolteachers. A number of public schools for girls between the ages of twelve and eighteen appeared that prepared girls for various teaching degrees; the range of schooling options for girls expanded, including for those who wished to pursue postprimary studies.

The law of June 17, 1881 represented a distinct step forward with the creation of at least fifty *écoles moyennes* (middle schools) for girls, staffed by a teaching personnel holding the official degree for *régentes*. The law encouraged towns and provinces to open their own *écoles moyennes* and to attach to these schools higher-level classes that prefigured the emergence of *lycées*. Above all, the program in these schools was aligned with that of boys' schools, although a few minor changes were made in order to introduce what were considered indispensable "feminine subjects": domestic economy and sewing. The latter only took up two hours per week in a program of twenty-eight hours (7 percent). Compared to boys, girls learned English at a younger age, beginning in the second year, but this was offset by simplified geometry and chemistry limited to "the applicable notions for domestic use." Finally, natural sciences were linked with a course in hygiene.

Despite the growth of girls' *écoles moyennes*, students attending such schools constituted a distinct minority in their age group. Increasingly, however, scholarships and free tuition encouraged some democratization. In 1884, of all the students enrolled in twenty-eight official *écoles moyennes* (the preparatory and middle section), 43 percent paid no tuition fees, and this meritocratic practice developed and became institutionalized after the war with the creation of the *Fonds des Mieux-doués* (foundation for the more gifted), which granted loans for secondary studies.[19]

By 1884, the Catholic party was once more in power and remained so until 1914. During this long "Catholic reign," the lower-level secondary system for girls suffered, the number of schools ceased growing, the programs became less ambitious, and subjects considered unnecessary for girls—geometry and natural sciences—were cut. In 1897, a new program eliminated the obligation to recruit teachers with State diplomas, thus allowing women religious to teach without official teaching qualifications. This same ruling once again tied girls' education to their roles as wives and mothers, aiming to inculcate them with: "feminine moral qualities and those of the good housewife [...] order and cleanliness, activity, exactitude, vigilance, thriftiness, good habits and economy."[20]

Special women inspectors were appointed to oversee the choice of reading matter and writing-exercises were expected to inspire "a love of the family and domestic virtues, to teach women's duties [...] and to bring to light the grandeur of mothers in the education of their children."[21] The program in mathematics was lightened and oriented "to deal with household matters, the objects of domestic economy, and the results of savings." Chemistry was associated with "culinary sciences" and drawing with sewing. Even the association Féminisme chrétien (1902), which supported the development of stronger studies for girls, warned these studies should remain within the limits of the "sacred role" assigned to women by nature and providence.

These regressive measures reflected the fears of Catholic elites following the social movements of 1886; they sought to restructure Belgian society around ideals of class collaboration and the complementarity of the sexes. In the process, they reaffirmed women's maternal and familial vocation as an essential component of the social order. And, beginning in 1889, they began serious efforts to develop a network of home economics schools that included a secondary level to offer girls domestic and familial training.

The political struggles that so marked girls' secondary education encouraged the development of schools but not necessarily female emancipation. Still, the progress of municipal autonomy helped girls' status, as it allowed the administrations of major cities to preserve their most advanced schools. On the eve of the First World War, a momentum clearly existed: girls were able to attend secondary schools, some of high quality, and, as of 1880, a small number of graduates of these schools had even moved into the universities.

The Interwar Period: Institution Building and Diversification

The public and private network of schools expanded rapidly after 1918, thanks to a new demand for schooling in an economy whose service sector was expanding, and to the decline in incomes during the war, which pushed many young bourgeois women to seek work. The impetus for development came once again from both Flemish and Walloon municipalities and the private sector: in 1915, the Institut Warocqué in Morlanwelz opened a section of modern humanities for girls; in 1917, the city of Schaerbeek (region of Brussels) created the first *lycée* for girls, complete with a section in the classical humanities. Between 1918 and 1925 eight municipal *lycées* emerged. The State moved more slowly to create a complete humanities program for girls, despite the pressure exerted by the *Fédération belge des femmes universitaires* (1921) (Belgian Federation of University Women). In 1925, the government adopted two strategies: it transformed certain *écoles moyennes* into *lycées* and, in places where the former did not exist, it opened boys' *athénées* to girls. This represented the beginning of a form of coeducation, but above all it created a humanities diploma for girls that gave direct access into the university without having to take a test. As a result, the numbers of *lycéennes* increased dramatically during the 1930s, and many moved on to the university: in 1936, of all students in humanities, 15 percent were women.

The opening of the *athénées* to girls generated gender tensions within the teaching personnel, which did not exist in the *lycées* where the staff was entirely female until 1970. The *Fédération belge* had won the government's promise that the percentage of male and female teachers would be proportional to that of boy and girl students, but this measure was never respected, prompting numerous recriminations.[22] As late as 1955, professional women denounced the paucity of women teachers in the coeducational *athénées*, while women teachers complained they were overburdened with administrative tasks compared to their male colleagues and often shunted aside for promotion.[23]

Meanwhile, the competition between schooling networks pushed Catholic institutes to bolster their academic programs, particularly once the Catholic university in Louvain opened its doors to women in 1920. Notable differences remained between this private sector and the secular public schools, less in the subjects studied than in the spirit and objectives of these studies. While *lycées* openly prepared their students for higher education, Catholic institutions continued to perceive these studies as "a preparation for a wifely and maternal vocation." A survey among the mother superiors of these institutes in the 1930s showed their commitment to a vision of secondary education that allowed women to fulfill their familial duties better, and to become "modern" wives and mothers.[24] The final certificate was not expected to allow students to work, except in cases of dire need. Indeed, the Filles de la Croix in

Liège argued that when girls pushed their studies too far they were less likely to find husbands; as a result, such studies should be reserved for a small number of women destined to remain spinsters.[25]

The interwar period reveals a number of paradoxical features. On the one hand, the demand for secondary education for girls expanded and their entry into university studies contributed toward raising the level of secondary programs. On the other hand, however, technical and professional orientations became increasingly segregated by sex.[26] Girls were oriented toward a limited range of "feminine" sectors (sewing and textiles) that corresponded to traditional conceptions about women's skills, despite the decline of the textile industry.[27] The opening of new sectors, allowing a greater range of opportunities for girls, nonetheless remained very marked by gender stereotypes. Belief in the superior intellectual abilities of men and their particular affinity to "science" remained dominant. As a result, the general rise in the level of girls' studies did not directly contribute to their (economic) emancipation.

Nonetheless, for the conservative defenders of women in the home, the growth of young women in secondary education generated anxiety. Various efforts were made in 1926, in 1930, and again in 1932 to slow down this evolution by requiring all girls to study home economics for two years prior to secondary studies. Proponents filed two proposals along these lines in the Chamber and the Senate,[28] generating vehement protest from feminists against these measures judged "most absurd, most vexatious and most arbitrary."[29] But the idea that domestic economy, hygiene, and puericulture constituted feminine knowledge par excellence persisted and provoked debate once again in 1937, with the proposal to create a complete secondary-level program in "women's sciences," followed by a university program.

Post-Second World War: Massification and Coeducation

After the Second World War, mass secondary education developed, including for girls. This was accompanied by important changes in women's civil and civic status: suffrage in 1948 and the abolition of the principle of husband's authority in 1958. In a world where knowledge and advanced schooling were increasingly indispensable, the State voted tuition-free secondary education in 1958; inevitably pupils spent more and more years studying.[30] The law of 1983 recognized this reality by declaring schooling obligatory until the age of eighteen.

Despite these democratic measures, gender differences and discriminations remained. In the 1950s the call for the creation of "familial humanities" in *lycée* studies reflected the conviction that girls also needed specifically feminine training. Within the technical secondary and professional sectors certain feminine niches emerged, such as that of home help; efforts to diminish the gender segregation in these areas achieved little.

After the Second World War, pedagogical reformers saw coeducation as the means to diminish differences within schools, but their project concerned class more than gender. In public schools, coeducation progressed spontaneously in the postwar period: 50 percent of secondary schools were coeducational in 1947 and 67 percent in 1953–1954. In Catholic schools, however, the resistance to coeducation remained strong due to the Pope's explicit refusal of all forms of coeducation: 0.2 percent of Catholic secondary schools were mixed-sex in 1964–1965.

Rendered obligatory at all levels of public schooling as of 1970, coeducation reflected a new social reality where women were increasingly present in the work force. As a result, the program for secondary studies applied to both sexes with only minimal differences between the two (such as the course in physical education, which remained single-sex). As of 1978, all secondary studies were decreed open to both sexes, but the private Catholic network, which was particularly influential in Flanders, opposed the decree and only instigated coeducation systematically in the 1990s.

The increasing feminization of the teaching staff has accompanied the development of secondary education: within the public sector, 56 percent of all professors were women in 1988, and 66 percent in 1998. This situation testified to the growth of women university graduates, but also men's gradual desertion of a profession whose prestige was sinking.[31] Upper-level positions of responsibility, however, have become increasingly masculinized; not only do men preserve positions that were masculine in the past but they also take over the directorships of previously feminine *lycées* (table 8.1).[32] The tendency is even stronger within Catholic schools: the disappearance of women religious as heads of schools was often followed by the appointment of a male lay director.[33] This masculinization is even stronger in the inspector corps, where the first women inspector was only appointed in 1974; until then male inspectors passed judgment on both boys and girls' schools.

Coeducation was the product, nonetheless, of a neo-feminist movement as well as institutional feminism beginning in the 1970s. It owes much to European directives that have stigmatized single-sex education as a form of discrimination and "apartheid." As elsewhere in Europe, coeducation was not the result of a thoughtful egalitarian pedagogical project, and, as a result, it quickly generated disappointment. Many had hoped that the end of spatial separations between boys and girls and the commingling of their studies would suffice to produce greater equality. The persistent gender preferences among subjects forces us today to rethink this process.

Table 8.1   The "glass ceiling" in secondary teaching

Responsibility	Year	Percentage of Women	Year	Percentage of Women
Wallonia + Brussels				
Head of school	1995–96	25.8	2000–2001	26.4
Flanders				
Head of school	January 1990	19.6	January 1996	18.2

Source: http://www. statistiques.cfwb.be; http://www.ond.vlaanderen.be

Notes

*Translated by Rebecca Rogers

1. Language was a particular concern in higher education, where French dominated until 1930 (opening of University of Ghent). Beginning in 1932, secondary education was offered in Flemish in Flanders, French in Wallonia, and in both languages in Brussels. The implications of these measures have not been studied from a gendered perspective.
2. See Eliane Gubin, "Libéralisme, féminisme et enseignement des filles en Belgique," in *Politique, Imaginaire et Education*, ed. F. Maerten, J.-P. Nandrin, L. Van Ypersele, *Cahiers du CRHIDI*, 13–14, (Bruxelles : Facultés universitaires Saint-Louis, 2000), 151–174.
3. This refers to teachers at the lower secondary level in charge of pupils between the ages of 12 and 14.
4. The State could intervene only when private initiatives were insufficient. Liberals defended the opposing principle, the State's duty to provide teaching; this debate was at the heart of the ensuing tensions with the Catholic Church. The issue was only resolved in 1958 (the "School Pact") when the State assumed the expenses of the entire secondary education network and exerted its inspection rights.
5. A. Tihon, "Les religieuses en Belgique du XVIIIe au XXe siècle : Approche statistique," *Revue Belge d'Histoire Contemporaine* 7 (1976): 1–54; Paul Wynants, "Les religieuses de vie active en Belgique et aux Pays-Bas, XIXe–XXe siècles," *Revue d'Histoire Ecclésiastique* 95 (2000): 238–256.
6. These schools represented upper secondary education for boys between the ages of fifteen and seventeen.
7. Valerie Piette, "Un réseau privé d'éducation des filles. Institutrices et pensionnats à Bruxelles 1830–1860," *Sextant* 13–14 (2000): 149–177.
8. Advertisements can be found in Archives de la Ville de Bruxelles, Fonds Instruction publique, n°63.
9. For more information on the women cited in this chapter, see Eliane Gubin, Catherine Jacques, Valérie Piette, and J. Puissant, eds, *Dictionnaire des femmes belges, XIXe–XXe s.* (Brussels: Racine, 2005).
10. *Revue encyclopédique* (Paris, 1832), 599.
11. Now forgotten, David-Eugène Lévi-Alvarès (Bordeaux 1794–1870) achieved notoriety in France and even Europe, thanks to his maternal educational courses in Paris (1821) and then his normal course for girls (1833). With his son Théodore, he applied a global method inspired by Rousseau; he organized teaching outside the classroom (concerts, museum visits) and worked closely with mothers (*L'Education de la Femme*, avril 1864); Eliane Gubin and Valérie Piette, *Isabelle Gatti de Gamond 1839–1905. La passion d'enseigner* (Brussels: GIEF ULB, 2004), 34–35, 39.
12. *L'Éducation de la Femme*, March 1862, 9.
13. *Cahiers féministes*, April 15, 1903.
14. The law of 1866 created two State normal schools for girls, but the first school only opened in Liège in 1874.
15. Eliane Gubin and Valerie Piette, *Emma, Louise, Marie... L'université libre de Bruxelles et l'émancipation des femmes* (Brussels: Ed. Arch. ULB-GIEF, 2004), 62–65.
16. Beginning in 1880 women students were able to enroll in the Université libre de Bruxelles, the same was true for the State universities of Liège and Ghent in 1881 and 1882, and finally in 1920 at the Université catholique de Louvain: A. Despy-Meyer, "Les étudiantes dans les universités belges de 1880 à 1941," *Perspectives universitaires* 3, 1/2 (1986): 17–49.
17. *Exposé de la situation du royaume 1876–1900*, II (Brussels, 1912), 466.
18. The liberals created the Ministry of Public Instruction in 1878; in 1884 the Catholics abolished it. From then until 1918, the administration of public education was divided between the ministries of the Interior, Industry, and Agriculture.
19. Paul Wynants, "École et clivages XIXe–XXe s.," in *Histoire de l'enseignement en Belgique*, ed. D. Grootaers (Brussels: CRISP, 1998), 67–68.

20. *Moniteur belge, Journal officiel*, March 12, 1897.
21. Ibid.
22. Gubin and Piette, *Emma, Louise, Marie*, 140.
23. XXVe Semaine sociale universitaire, *La condition sociale de la femme* (Brussels: Institut de Sociologie Solvay, 1955), 78, 79, 81.
24. *La Cité chrétienne* (1930–1931): 448–449, 586, 735–739, 929–931, 1024.
25. *La Femme belge* (July–August 1933), 280. Belgian bishops only abolished the requirement of celibacy for lay personnel in Catholic schools in 1962. In other professions, the law only annulled marriage bans for working women in 1969.
26. Grootaers, *Histoire de l'enseignement en Belgique*, 406.
27. E. Plasky, "Avant-propos," *La protection et l'éducation de l'enfant du peuple en Belgique* (Brussels, 1928).
28. Annales parlementaires, Sénat, séance du 27 mars 1932; Documents parlementaires, session 1930–1931, n°76.
29. *Égalité*, March–April 1935.
30. See Grootaers, *Histoire de l'enseignement*, 88–103.
31. ONSS, *Nombre d'employés et employées recensés au 30 juin 1998 à la Sécurité sociale* (Brussels), 1998.
32. *L'école au féminin* (Brussels: Université des Femmes, 1991), 81–82.
33. *Nationale Vrouwenraad* 3 (1990).

Chapter 9

Lutheranism and Democracy: Scandinavia

Agneta Linné

The story of girls' secondary education in Scandinavia from the eighteenth to the twentieth century is a story of gender and social class. It is a story of Lutheran faith, explicitly emphasizing the different callings of men and women in the evangelical striving for a good and decent life. It is also a story of the recognition of enlightenment and democracy as dominant social values influencing ways of life. Throughout the story, we hear the voices of pioneer women, enacting entrepreneurship, making alliances, and balancing on boundaries between conflicting interests. Girls' secondary education opened doors for women to fulfill a professional career at times when woman's position was considered to be that of mother and housewife. Inquiring into girls' secondary education, therefore, means inquiring into the relations between public and private.

Research to date includes general inquiries[1] and in-depth studies of girls' education in particular historical periods,[2] statistical overviews of the number and character of early schools based on school records and other archival data, analyses of curricula and/or the variety of textbooks.[3] Studies of the changing norms that guided informal education for upper-class girls in the eighteenth century who were educated by a governess, or who attended a small private day school or boarding school, have focused inquiries on advice books, prescriptive literature, memoirs, diaries and letters.[4] Two key publications by Schånberg analyze the historical transformations of girls' secondary education between 1870 and 1970 in relation to shifting demands for women in the work force.[5]

Individual biographical studies of founders of particular schools,[6] collective biographies of a group of founders,[7] and in-depth research into the history of a school and its leading teachers[8] have provided insights into the content and form of girls' secondary education. Studies looking at the differences between girls' and boys' secondary education include the influential study by Florin and Johansson of culture, social class and gender in the Swedish grammar school 1850–1914,[9] and Johansson's inquiry into the meaning of being a student at girls' and boys' secondary schools in the twentieth century, for which she used interviews and life histories alongside various written school records.[10] Other research has focused on ways in which teacher education opened doors to secondary education for girls.[11] Historical overviews of the educational landscapes of the Scandinavian countries provide the context to understand developments in girls' education.[12]

Given that national boundaries in Scandinavia shifted dramatically over the period,[13] summarizing the growth and transformation of girls' secondary education in Scandinavia during 1700–2000 poses challenges of definition in geographical terms. The chapter focuses on girls' education in Sweden, Denmark, and Norway. Summarizing girls' education from 1700 also raises questions around the definition of "secondary" education, particularly in the early eighteenth century, when home-schooling was common for both boys and girls.[14] The chapter distinguishes four major phases in the history of girls' secondary education in Scandinavia: the period 1700 until about 1790, when the first steps were taken toward establishing more formal girls' schools on what might be considered a secondary level; the last decade of the eighteenth century to about 1865, when girls' secondary schools became established in most cities on private market terms with a separate curriculum; further development toward an integrated secondary education from 1870 until around 1930; and the period to 2000, with integration into the public coeducational school system.

From Homeschooling to Daughters' Schooling: Toward Establishing a Secondary Education for Girls

In the eighteenth century the French influence on the upper classes of Northern Europe was considerable. The education of the daughters of the nobility combined homeschooling and broader cultural refinement. In addition to reading, writing, arithmetic, and needlework, girls learned French, history, and geography, together with dancing, music, fine arts, letter writing, and the art of polite conversation. The aesthetic subjects were important markers of social and cultural capital. The moral ideals deemed suitable for a girl of the aristocracy permeated literature for young women.[15] The honorable young boy and girl were to strive continually for greater perfection as Christians. Modesty but not timidity was a desirable goal for daughters. The Scandinavian women active as writers and translators from German, French, and English between 1720 and 1772 acquired elaborate writing skills and competence in foreign languages that demonstrates what can be considered a secondary education.[16]

As the influence of France became increasingly dominant in Scandinavia during the eighteenth century, it also became important for upper-middle-class girls outside the aristocracy to learn French. Since neither French nor other living foreign languages were taught in the public schools, this increased the market for private teaching. In the larger Swedish cities, several minor day schools or boarding schools were set up, some of which admitted girls. Boarding schools were not founded as frequently in Scandinavia as in other European countries, however. The secondary schools that were established from the 1780s were mainly organized as day schools, a pattern that continued in Scandinavia. In 1913, a Swedish survey of secondary schools admitting girls documented eighty-one girls' schools and nine coeducational schools, but only three boarding schools, of which only one had more than twenty pupils.[17]

In a situation where public schools were few and poorly accepted by the upper social classes, homeschooling formed part of the educational market. Hammar has shown the considerable use of homeschooling compared with public schooling in Stockholm in the first half of the eighteenth century.[18] Although documentation is scarce, letters, diaries and memoirs for the first three-quarter of the eighteenth century reveal the presence of French governesses in some families. The most commonly used textbooks and manuals of French grammar indicate that girls and young women were among the imagined audience and that grammar and letter-writing skills were taught. Recommended reading matter included Fénelon's *Télémaque*, the *Magasin des enfants* by Jeanne-Marie Leprince de Beaumont, as well as works by Étienne Chouffin, Madame de Genlis, Madame d'Épinay, and Charles Rollin.[19]

During the latter half of the eighteenth century, the growing debate in Denmark-Norway and Sweden about the education of middle-class girls went hand-in-hand with the publication of advice books and other educational texts. Alongside increasing concern with matters like rationality and love in relation to marriage was discussion about the suitability of scientific subjects for girls' education.[20] Changing economic realities meant that not all middle-class girls could look forward to life as a married woman. As a result, providing for daughters became a problem for many bourgeois parents who developed new expectations about girls' education. Enlightenment ideals together with a radical evangelical revival, including pietism and philanthropy, challenged upper- and middle-class fathers to take an interest in their daughters' education.

Jean-Jacques Rousseau's *Emile* (1762) and *Julie ou La nouvelle Héloïse* (1761) offered models for women's new, gender-specific moral obligations and their "proper" place in life. These novels warned girls of the dangers of an in-depth education and of becoming an intellectual.[21] The German pedagogues Basedow, Salzmann, and Campe were among the early reformers who influenced the Scandinavian debate.[22] A Lutheran revival movement, the Herrnhuthian Evangelic Brother Society, founded the first Swedish girls' secondary day school (*Societetsskolan för döttrar*, the Society School for Daughters)[23] in Gothenburg in 1786. The Herrnhuthian revival focused on a personal relationship between Christ and the individual and advocated a form of equality between men and women, despite their different callings in life. Although their membership was small, and religious meetings outside the state church became prohibited, the Pietist and Herrnhuthian movements spread within bourgeois circles in the capitals and larger Scandinavian cities. Gradually, their ideals, as well as notions of religious freedom and individual rights, spread within the overarching Lutheran state church.[24] Another, and more radical, influence can be traced to the ideas of the writer Mary Wollstonecraft. Wollstonecraft journeyed to Sweden, Norway, and Denmark in 1795. By this point she had published *Thoughts on the Education of Daughters* (1787) and *A Vindication of the Rights of Women* (1792), in which she defended equal rights between women and men. In the travel book of her

Scandinavian journey, published at the turn of the year 1795–1796 and translated into Swedish in 1798, she refers to the subordinate status of Scandinavian women. While the travel book became a great success, the effect of her visit is difficult to evaluate. One imagines, however, that her ideas circulated among the influential commercial and merchant classes who hosted her, and may have reached the Danish Prime Minister Struense, who received her in an audience.[25]

The Danish Home Office conducted a survey of schools in Denmark and Norway in 1801. This revealed the existence of thirty-one private schools for girls and fourteen coeducational schools.[26] How many of these schools were at a secondary level was not reported, although it is likely that the coeducational schools were elementary. In Scandinavia, public elementary schools were gradually established from 1800 onward and usually catered for both boys and girls (often taught separately at schools in larger cities). The first Scandinavian coeducational school at an upper secondary level was the private *Palmgrenska samskolan* (Palmgren coeducational school). Founded in 1876 in Stockholm as a practical elementary school, it had developed by 1886 into a lower and upper secondary school.[27]

Two of the best-known Danish "real" schools for boys opened in 1786 and 1787. The term "real" school (German *Realschule*) applied to schools above the primary level, teaching scientific and civic subjects and one or more foreign languages (excluding Latin and the classics) to children who could already read.[28] Nonprofit associations ran the two Danish "real" schools for boys and prepared their pupils for future tasks in trade and commerce. In 1787, parents published two proposals for "real" schools for girls, one of which opened for girls aged 6–15 that same year. The curriculum included so-called ladies' accomplishments (mainly needlework as well as housekeeping), alongside academic subjects. Within a few years, however, parents became critical of the school's governance and its limited scientific and professionally useful subjects. A group of parents established a new day school in 1791 that became known as *Døttreskolen* (the Daughters' school). By 1801 it had developed into a woman's academy, teaching Danish, French, German, religion, morals, writing, arithmetic, natural history, geography, history, reading comprehension, needlework (knitting, linen sewing, embroidery, and dressmaking), drawing, and, as electives, dancing and English.[29]

In Norway, a lower secondary day school, *Trondheim's borgerlige Realskole*, opened in 1783. This admitted both boys and girls, who were taught separately and offered different curricula, with girls studying at the special daughters' school (*døttreskole*). The subjects taught to both girls and boys included religion, recitation, history, geography, mathematics, writing, and drawing. In addition, girls were taught techniques thought particularly fitting for women, while boys were trained in commerce, crafts, and navigation.[30]

Although some egalitarian voices were heard in public, the syllabi and content of the first secondary schools for girls reflected deeply rooted Lutheran ideals in Scandinavian eighteenth- and nineteenth-century culture and mentality about women's and men's separate missions in life. Nonetheless, as Gold argues, in practice there was much ambivalence. While manuals for girls and women proposed school curricula that stressed the virtues of piety, purity, domesticity, and submissiveness, utilitarian ideas about practical and useful knowledge were reflected in the voices of the founding fathers who opened private girls' schools. In these schools, many parents chose a more "scientific" education for their daughters. Studies in learned and intellectual subjects like Latin, the classics, and advanced mathematics were seen as subjects for boys only and parents advocated foreign languages, the mother tongue, and general and practical subjects for girls. Given boys' and girls' different callings, however, Scandinavian society preferred single-sex schooling above the elementary level. In this fashion, institutions emerged that trained good, enlightened mothers and wives, while also preparing daughters for practical tasks in an expanding labor market.

Educating Mothers, Wives, or Citizens: The Founding Period (1790s–1860s)

Between the latter half of the eighteenth century and the mid-1860s, formal girls' education at a secondary level emerged in all the Scandinavian countries.[31] While the State supported secondary education for boys, it left the establishment of girls' secondary schools to a dynamic private sector that responded to an increasing demand for education from the middle classes. Demographic concerns played into this growth, as women outnumbered men in Scandinavia during the nineteenth century and marriage frequency was low.[32]

Although early records of girls' education are incomplete, Schånberg documents the creation of around thirty more permanent girls' schools in the larger Swedish cities before 1865. The Herrnhuthian Evangelic Brother Society, mentioned above, founded the first of these schools (in Gothenburg 1786); next came schools in Askersund (1812), Gothenburg (1819), Stockholm (1831), and Gothenburg (1836).[33] These schools differed from the older type of more academic schools in having formal syllabi and modern educational methods. They also differed from older charity girls' schools by including foreign languages in the curriculum. Gold documents forty-three schoolkeepers running female "real" schools or academies between 1790 and 1817 in the Danish capital Copenhagen.[34] Ytreberg charts the gradual establishment of more academic girls' schools in the Norwegian capital Kristiania (now Oslo). By the 1820s there were five girls' schools in Kristiania, and by 1837 schools similar to the *Realskole* in Trondheim had been founded in nineteen Norwegian cities.[35] Upper-class families still used home teaching, however, and sent their girls for periods of study at schools in Copenhagen or other Danish cities. The well-known Norwegian writer Camilla Collett, born Wergeland (1813–1895), for example, received a good education, first at the vicarage in Eidsvoll, and later at the Herrnhuthian school in Christiansfeld in Schleswig.[36]

In 1831, the bishop Johan Olof Wallin and the influential historian and pedagogue Anders Fryxell established the first modern girls' secondary school in the Swedish capital of Stockholm, the *Wallinska skolan* (the Wallin school).[37] A similarly modern institution, the *Nissens pikeskole* (Nissen's girls' school) was founded in 1849 in Kristiania in Norway.[38] The two schools shared many similarities, reflecting the circulation of ideas about girls' secondary education in Scandinavia.[39] Both sets of founders drew on Lutheran views of boys' and girls' separate natures and missions in life, which they thought required a separate secondary school curriculum. This was more academically demanding, however, than the earlier schools and included foreign languages. Both institutions sought to develop girls' practical sense by offering knowledge that would prove useful later in life. Since women's mission was to support men as their companions, and to raise the future generation, the founders argued women's education should enable women to understand the higher meaning of life and should instill pious and noble feelings, while developing a clear mind to allow women to govern their feelings. Above all, the educational ethos emphasized Christian belief and hope that would allow women to direct their children toward God. Within these schools, religious values, rather than dogmatic teachings, permeated the entire life of the school. All academic subjects were to serve to strengthen religious consciousness. In history lessons, students learned how Christianity introduced a new light in the world. However, history as a school subject had value of its own within girls' schools as a substitute for the experiences of boys in their future civic or public life. In spite of the founders' view that wide-ranging and intellectually demanding studies were suitable for boys only, these modern schools promoted women's right and duty to seek real educational cultivation and defended a vision of women's nature that justified the teaching of a serious, rather than superficial, version of history and other scientific subject matter.[40]

Nissen was inspired by the Danish priest, writer, and pedagogue N.F.S. Grundtvig (1783–1872), father of the Scandinavian *folkhögskola* (folk high school, a residential college for adult education). Typical of his time, Nissen advocated that the curriculum should promote civic and national spirit through the study of literature, nature studies, geography, and foreign languages. He thought that practical subjects like writing and arithmetic were indispensable to both men and women. For the latter he advocated a solid education in music and arts in order to develop the sense of beauty that was invaluable to woman in her duty as mother of the household. Physical education was also judged important to free female pupils from affectation and constraint. Nissen's academic goal was to achieve the harmonious development of the intellectual human being, while the school's wider aim was to become an institution for educational cultivation to foster the development of both mind and character.

On opening in 1849, Nissen's school started with the primary level and only twenty-three pupils. By 1852 it covered ten school years with 122 pupils and offered three foreign languages (German, English, French) in the upper grades. Nissen's school became a model for other girls' schools in Kristiania. An important step was taken in the 1860s when the school created Norway's first special course for women teachers, reinforcing its status as a model school. From 1862, the course for women teachers consisted of one year of general subjects and a second year more focused on pedagogical and educational matters. The course was later extended to three years. Until the University of Oslo admitted women students in 1882, Nissen's course provided the most advanced education for women teachers.[41]

In Denmark, Nathalie Zahle (1827–1913) became an outstanding pioneer of girls' education. Her father was a priest and she received her primary schooling at the vicarage. Both her parents died when she was nine. Between nine and fifteen, she was a pupil at the Copenhagen Døttreskolen (founded in 1791), where she obtained a reasonably good secondary education, considering what was available at the time. After six years working as a private teacher, including five years in the home of chief physician N.E. Holst, Zahle spent a year studying at a private institute, which prepared women for the examination required for the management of a private school, which she passed with the highest grade.[42] Zahle's first plan for a girls' school in 1851 contained ideas similar to those of Nissen, but advocated more similarity with the education of boys. Zahle established her school in 1852 with fifty-five pupils. By 1860 the number had risen to 202 and the school had become the largest girls' school in Denmark and included a special course for women teachers. Zahle's educational ideals combined traditional female and male virtues—purity was associated with discipline and industriousness.

Nathalie Zahle linked her school project with the development of opportunities for daughters and women. She focused on groups between the aristocracy and the working-class. However, she widened girls' education to include industriousness, entrepreneurship, strength of will, and creativity in public society as well as in the private home. Nonetheless, her educational project was ambiguous. Daughters were expected to develop an aesthetically attractive and mild appearance in line with notions of femininity, although she criticized shyness, timidity, and uncertainty. Zahle used her influence to widen the opportunities for her students, while acting as a dominating figure in their school life.[43]

Intense debates about schooling, including girls' education, took place inside and outside the Swedish parliament in the 1840s and 1850s. The men behind the Wallin School were important participants in the parliamentary debates. In her influential novel *Hertha* (1856), the internationally renowned writer Fredrika Bremer (1801–1865) spoke strongly in favor of women's right to education. Influential liberals argued that women would be useful as elementary school teachers, mindful that employing women teachers would be cheaper. By the 1840s and 1850s, a few women had been admitted informally as students to the Stockholm Elementary School Teachers' Seminary. Parliamentary and governmental decisions in 1859 and 1864 converted two elementary teachers' seminaries into women's institutions. An advanced seminary for women teachers was founded in Stockholm in 1861, the Royal Advanced Teacher Seminary for Women (*Kungliga Högre Lärarinneseminariet*). This created a route for women to acquire the formal qualifications to teach at the elementary teacher training seminaries for women and at the girls' schools. Fredrika Bremer took a special interest in this female academy, considering it a fruit of her novel *Hertha*. Both the advanced and the elementary teacher seminaries were state institutions.[44]

In 1845 in Denmark, the Copenhagen school authorities established an examination for those who wished to direct a school, and in 1859 women were entitled to undergo the elementary school teachers' examination. Unlike in Sweden, aspiring teachers had to prepare at a private seminary or on their own, since a state seminary for women teachers was only established in 1918.[45] In Norway, women formally achieved the right to become elementary school teachers in the 1860s, and by 1890 were entitled to study at all public teacher seminaries.[46] Granting women the right to teach at the primary school level, where they predominated in the very first school grades, opened an important sector of the labor market to women.

The Royal Advanced Teacher Seminary for Women in Sweden came to be of great importance in setting norms for the curriculum and teaching methods in girls' schools. It also provided a most important road for women to achieve an upper secondary education. The principal and the teachers of theoretical subjects were male professors, often with a university doctorate, whereas a woman teacher served as deputy headmistress. The seminary offered instruction in general subjects such as religion, the mother tongue, literature, French, German and English, history, geography, mathematics, biology and the natural sciences, alongside lessons in the arts, music, physical education, and the pedagogical and didactical aspects of teaching. The general course lasted three years, with an optional fourth year. The seminary served as a woman's academy for generations of women excluded from the public secondary schools that only admitted boys. A "model school" soon became associated with the seminary to offer practical experience for the seminary students; later its curriculum acted as a model for girls' schools that aspired to public financial resources.[47] As it was a state seminary, costs were reasonable both for the seminary course and the model school. This contrasted with the private girls' secondary schools that excluded girls whose parents could not afford the fees. In addition, the elementary school teacher seminaries open to women offered at least a lower secondary education, although they did not include a foreign language until the beginning of the twentieth century. In many cases, women who acquired formal positions as teacher educators made remarkable contributions to modern public debates about schooling, the child, education, and philanthropy. The dominant actors in the pedagogical arena made teacher education an important site in the gendered struggles around modernity.[48]

Expansion, Consolidation, and Municipal Schools: Steps toward Integration (1870s–1930)

The end of the nineteenth and beginning of the twentieth century was a time of great social transformation in Scandinavia as boundaries between the private and the public and between State and civil society shifted. Feminists, politicians, and intellectuals all debated intensely the question of universal suffrage and the place of women. Within this context, the development of secondary education for girls stimulated a range of strategies.[49]

The special girls' school commission of 1866–1868 suggested to the Swedish parliament that the State should finance some secondary schools for girls in the same way as it financed boys' grammar schools in a number of cities. Because the tuition fees of the existing girls' schools were high, even for the bourgeoisie, many liberal fathers favored the proposal. The key arguments in the parliamentary debate of 1873–1874 revolved around the responsibility of the State to finance only schools that gave secondary education to prepare future upper-level state officials—and, according to the law, only men were entitled to hold such offices. Schools educating future clerks and lower civil servants as well as artisans and craftsmen were the responsibility of municipalities, private organizations, or citizens. The primary responsibility of the State in relation to the education of girls was thought to be teacher education. Nonetheless, the parliamentary majority decided to give a government subsidy to girls' secondary schools that fulfilled certain conditions. These schools became subject to government control, had to submit annual reports, and were required to offer some free places to pupils of limited means. The women's movement was disappointed with the outcome.[50] Still this decision brought public recognition of the girls' secondary schools and helped in the process of consolidation.

Gradually, a few exceptional schools in the larger cities began to compete for the right to hold the academic upper secondary school certificate (*mogenhetsexamen*, matriculation, equivalent to the French *baccalauréat* and the German *Abitur*) that would open the doors of the university to successful girls. Access to higher education in Scandinavia had traditionally been achieved through an entrance examination held by the university at their campus (only men were accepted as university students). In Sweden, this examination was moved to the upper secondary grammar school in 1864 and was organized as a national, state-controlled examination upon leaving the last grade. Only boys were initially entitled to take part.[51] In Norway, the exam (*Artium*) was moved from the university to the grammar schools in 1883.[52] In Sweden, girls gained formal access to the examination in 1870 (the first candidates, however, had to register for the exam on a private basis), and the first woman was accepted as a university student in 1873. In 1874, the Wallin school in Stockholm gained the right to hold the examination.[53] In Denmark, women gained access to the academic upper secondary school certificate and achieved the right to study at the university in 1875,[54] and in Norway in 1882.[55] In 1877, the first two Danish women passed the academic upper secondary school examination as private students.[56] Zahle's girls' school achieved the right to hold the academic upper secondary school certificate examination in 1886.[57] Gaining the right to hold this examination was a demanding task that required years of preparation for the more ambitious schools. In 1900, only four Swedish girls' schools were entitled to hold the examination. By 1913, ten girls' schools and seven coeducational schools out of a total of eighty-one higher girls' schools and nine higher coeducational schools in Sweden could do so, of which ten were situated in Stockholm.[58] The examination rights, however, were double-edged, as they restricted considerably the freedom of the schools to innovate in the curriculum.

State subsidies for girls' secondary schools contributed to the expansion of schools that followed. Between 1865 and 1885, eighty-five new girls' schools were founded in Sweden, bringing the number of girls' secondary schools to somewhere between 120 and 124. From the mid-1870s, regular girls' school meetings formed an arena where girls' education could be discussed in public. Some girls' school curricula increasingly imitated the curricula of boys' grammar schools, although important gender differences remained, including the presence of needlework: girls did not study the classical languages and often learned more French than German, whereas the grammar schools devoted more hours to German (in addition, Latin and Greek). Girls studying at the secondary schools acquired a broad, modern education that was attractive to wide circles in the expanding cities, but they remained fee-paying in contrast to the boys' state grammar schools.[59]

The ambiguities of girls' secondary education deepened in the late nineteenth century. The transformations at the turn of the century in the labor market, economy, population, city growth, means of communication, and ideas diversified standpoints when it came to the education of women. The right of some schools to hold the academic upper secondary certificate examination, and the fact that male grammar school teachers occasionally taught at the girls' schools, drew these schools closer to the curricula of the traditional boys' grammar school.[60] The first woman in Sweden to achieve a Ph.D., Ellen Fries (in 1883), was employed at one of the larger girls' schools in Stockholm, with the task of organizing a special upper secondary school division. She thought girls should have the same right to higher education as boys and based her argument on the egalitarian idea that the two sexes are of the same human race, in the first instance. In principle, she thought that girls' schools contributed to conserving differences between the sexes and should be reformed; however, she later spoke against her earlier ideals.[61] Integration and differentiation ran side by side. Most Swedish girls' schools offered an academic track without formal examination, which many thought was better suited to woman's nature and future mission in life. The tension between educating girls for the home, or for citizenship and the labor market, continued over the coming years. Strong voices in Scandinavia advocated the girls' school as a better alternative for girls than taking the academic upper secondary certificate. They also worked for the inclusion of new, practical subjects like home economics in the curriculum of the girls' secondary schools, and they frequently referred to woman's fragile biology, arguing it made her unfit for intellectual studies or male competition.[62]

The 1880s and 1890s was a period of reform around new ways of teaching and child-centered education, influenced by psychology as a new and promising science and by German school experiments, in which Scandinavian women school pioneers played their part. Nordic school meetings, the development of Grundtvig's educational ideas and the Danish folk high schools, Otto Salomon's craft courses at Nääs, and the pedagogy developed in girls' schools like the one founded by Anna Sandström (1854–1931) attracted great interest among large numbers of teachers.[63] The intimate, homely atmosphere of the girls' school, with its emphasis on humanistic subjects and foreign languages, optional courses, and a personal way of teaching religion, did not easily accommodate to the centralized control of the academic upper secondary examination, especially when these schools also operated a private teacher education seminary, as did the schools of Anna Sandström and Nathalie Zahle.[64] The private, coeducational school of Sofi Almquist, for example, officially founded in 1892, continually experienced difficulties with examination requirements and central regulations.[65]

In 1892, a proposal to start an association to support a coeducational school was published in the Stockholm newspapers. Its director, Anna Whitlock (1852–1930), later to become a leader in the women's suffrage campaign, was deeply critical of the public elementary and secondary schools for what she saw as their dogmatic and routine teaching, particularly of religion. In 1878 she had started a small school with only seven pupils at the home of her friend, the writer Ellen Key,[66] who became her teacher colleague. The teachers were Anna Whitlock's friends, and the atmosphere of the school informal and homelike.[67] In 1892, Anna Whitlock proposed the foundation of a coeducational school informed by progressive child-centered educational methods. The coeducational school curriculum contained less formal study and more practical work and physical exercise. Whitlock offered a range of optional subjects, and subject content was to follow the child's mental development, stimulate pupil autonomy, develop the child's individual talents and aptitudes, and foster citizenship. A number of leading liberal intellectuals supported her proposal. Later that year, at a meeting held in the auditorium of the Academy of Science to promote the new school, a prominent representative of the state church emphasized the new way of teaching religion that was to characterize the school, and medical authorities spoke strongly in favor of science, hygiene, sound physical exercise, and collaboration between young boys and girls. Religion was optional and non-confessional in Whitlock's school, an extremely radical stance in early twentieth-century Scandinavia. Her school soon became one of the few girls' or coeducational schools that achieved the right to hold the academic upper secondary school examination.

In the expanding cities of Copenhagen, Stockholm, and Kristiania, demands for schools for girls, as well as schools practicing the "progressive" child-centered pedagogies, were increasing, not least from the growing upper middle classes. The entrepreneurial strategies of Nathalie Zahle, Anna Sandström, Anna Whitlock, Ragna Nielsen (1845–1924), who founded the first coeducational secondary school in Norway in 1885, and their fellow school founders, enlarged the space for women. Their strategies included risk-taking, substantial organizational capacity, endurance, the ability to negotiate, and to raise social and economic support in order to recruit pupils and finance the building of schoolhouses. Within the fast-growing Scandinavian capital cities, they filled new positions that were highly significant for the future shape of women's rights and women's location in the public sphere. In Scandinavia it is significant that the public space, into which many women struggled to gain entrance (not least in respect of female teachers), formed part of the State.[68]

In 1903, Danish girls gained access to the public lower and upper secondary schools, previously restricted to boys. In Norway a law of 1896 opened these schools for girls.[69] In 1905, Swedish girls entered the lower secondary level at some of the state grammar schools in smaller towns. Public middle schools for both girls and boys were established in the municipalities, in Norway by the 1880s and in Sweden from 1909, offering competition to the older private girls' schools, which had to struggle to survive. In 1918, the Nissen school was sold to the municipality of Oslo. In Sweden in the 1920s, however, the secondary school system was still highly segregated into schools for boys and schools for girls.[70] In 1930, around 20,000 girls studied at private or municipal girls' schools. They formed 55 percent of pupils at municipal middle schools, 24 percent of those attending lower secondary state schools, and 15 percent of the upper secondary state school population.[71]

Integration, Democracy, and Public Education—End of Story?

Following radically changing economic, social, and market conditions, and a growing ambiguity regarding the goals and content of girls' education, further steps were taken to integrate boys' and girls' secondary education in Scandinavia. A new generation of female school leaders with an academic education emerged.[72] Many private girls' schools were taken over by municipalities, and increasing numbers of students opted for municipal secondary schools. In Sweden, they also chose the upper secondary state schools that were eventually reorganized in 1927 to accept girl pupils.[73] The school systems of the Scandinavian countries were reformed to facilitate entrance to secondary schools for all categories of students. When government investigations in the 1940s were preparing for reform of the entire Swedish school system, many testified to the impressive achievements of girls' schools and their founders.[74] However, their days were numbered. The future belonged to coeducation and comprehensive education. In 1950, there were only forty-seven municipal and six private girls' schools still in existence in Sweden. In 1962, parliament voted to close the separate girls' schools.[75]

As part of school reforms after the Second World War, the new comprehensive school replaced the elementary school and the grammar school for the first nine years of schooling. The vision was for a democratic school to educate liberal individuals. These reforms have offered the same formal educational opportunities to girls and boys in Scandinavia over the past fifty years, with the upper secondary school open to girls and boys on equal terms. However, the selective system that still exists at the upper secondary and university levels complicates the picture.[76] At the universities, women still choose specialities that may be considered to extend their traditional role in the home (teachers, nurses, medical doctors, social workers). Formal equality between men and women has been achieved; but young men and women still choose different educational and professional futures.

The history of girls' secondary education covers a world of complexities. Some girls' schools developed as places where progressive child-centered pedagogies were tried and developed, others resembled the grammar schools for the male elite. Historical studies highlight that girls' schools equipped young women with the knowledge, self-confidence, and courage to take steps into the public sphere. By maintaining separate spaces, schools created a "room of their own" for young girls. That room furnished them with the capacity to accept challenges from the male professional arenas. Ambiguity remained a permanent presence in this process. Nonetheless, teacher education opened doors to significant networks, professional development, and future employment. Women school pioneers contributed in important ways to the changing shape of the educational field. Fighting for recognition and for political, educational, and human rights on equal terms with men created diversified opportunities and new positions for Scandinavian women.

Notes

1. Ebba Heckscher, *Några drag ur den svenska flickskolans historia* (Stockholm: Norstedts, 1914); Gunhild Kyle, *Svensk flickskola under 1800-talet* (Gothenburg: Kvinnohistoriskt arkiv, 1972).
2. Carol Gold, *Educating Middle Class Daughters: Private Girls Schools in Copenhagen 1790–1820* (Copenhagen: The Royal Library & Museum Tusculanum Press, 1996).

3. Gold, *Educating Middle Class Daughters*; Elisabet Hammar, *Franskundervisningen i Sverige fram till 1807* (Stockholm: Föreningen för svensk undervisningshistoria, 1981); Kyle, *Svensk flickskola*.

4. Eva Lis Bjurman, "Sophie, Education and Love: A Young Bourgeois Girl in Denmark in the 1790s," *Ethnologica Scandinavica* 26 (1996): 5–24; and *Catrines intressanta blekhet: Unga kvinnors möten med de nya kärlekskraven 1750–1830* (Stockholm/Stehag: Symposion, 1998); Hammar, *Franskundervisningen*; Ann Öhrberg, *Vittra fruntimmer: Författarroll och retorik hos frihetstidens kvinnliga författare* (Södertälje: Gidlunds, 2001); Jessica Parland-von Essen, *Behagets betydelser: Döttrarnas edukation i det sena 1700-talets adelskultur* (Södertälje: Gidlunds, 2005).

5. Ingela Schånberg, *Genus och utbildning: Ekonomisk-historiska studier i kvinnors utbildning ca 1870–1970* (Stockholm: Almqvist & Wiksell, 2001); and *De dubbla budskapen: Kvinnors bildning och utbildning i Sverige under 1800- och 1900-talen* (Lund: Studentlitteratur, 2004).

6. Birgitte Possing, *Viljens styrke: Nathalie Zahle—en biografi om dannelse, køn og magtfuldkommenhed* (Copenhagen: Gyldendal, 1992).

7. Annika Ullman, *Stiftarinnegenerationen: Sofi Almquist, Anna Sandström, Anna Ahlström* (Stockholm: Stockholmia, 2004); Agneta Linné, "Female Curriculum Pioneers, Modernity, and the Public Sphere around 1900" (Unpublished paper American Educational Research Association, 2006).

8. Nils A. Ytreberg, *Nissens pikeskole 1849–1949* (Oslo: Kirstes Boktrykkeri, 1949); Harald Thuen, "Hartvig Nissen," in *Pedagogiske profiler: Norsk utdanningstenkning fra Holberg til Hernes*, ed. Harald Thuen and Sveinung Vaage (Oslo: Abstrakt forlag, 2004), 65–80.

9. Christina Florin and Ulla Johansson, *"Där de härliga lagrarna gro...": Kultur, klass och kön i det svenska läroverket 1850–1914* (Kristianstad: Tiden, 1993).

10. Ulla Johansson, *Normalitet, kön och klass: Liv och lärande i svenska läroverk 1927–1960* (Stockholm: Nykopia, 2000).

11. Anna Sörensen, *Växelundervisningssällskapets normalskola och folkskoleseminariet i Stockholm 1830–1930* (Uppsala: Almqvist & Wiksell, 1930); Rudolf Fåhræus, *Högre Lärarinneseminariets historia* (Stockholm: Norstedts, 1943); Adda Hilden, *Fromme, stærke Kvinder: Lærerindeuddannelse 1800–1950, Dansk læreruddannelse 1791–1991*, vol. 3 (Odense Universitetsforlag, 1993); Agneta Linné, *Moralen, barnet eller vetenskapen? En studie av tradition och förändring i lärarutbildningen* (Stockholm: HLS Förlag, 1996); and "Morality, the Child, or Science? A Study of Tradition and Change in the Education of Elementary School Teachers in Sweden," *Journal of Curriculum Studies* 31 (1999): 429–447.

12. Gunnar Richardson, *Svensk utbildningshistoria* (Lund: Studentlitteratur, 2004); Wilhelm Sjöstrand, *Pedagogikens historia*, vols. 1–3:2 (Lund: Gleerups, 1961–1965); Gro Hagemann, *Skolefolk: Lærernes historie i Norge* (Oslo: Ad Notam Gyldendal, 1992); Elisabet Hammar, "Franskan," in *Frihetstiden: Signums svenska kulturhistoria*, vol. 4, ed. Jakob Christensson (Lund: Signum, 2006), 439–457; Matti Klinge, *Krig, kvinnor, konst* (Helsinki: Schildts, 1997); Adda Hilden and Anne-Mette Kruse, eds., *Pigernes skole* (Århus: KLIM, 1989); Jessica Parland-von Essen, "Flickors fostran," in *Gustavianska tiden: Signums svenska kulturhistoria*, vol. 5, ed. Jakob Christensson (Lund: Signum, 2007), 227–247.

13. In 1700, Sweden controlled parts of today's Baltic states and Northern Germany, Denmark ruled parts of today's Germany, and Norway was in a union with Denmark. By 1809, Sweden had lost control of all areas south and east of the Baltic and surrendered Finland (a province of Sweden since the Middle Ages) to Russia. In 1814, Denmark (which took part in the Napoleonic wars of 1807–1814 on the French side), was forced to give up the union with Norway; Norway entered a union with Sweden, which had fought on the other side, a union that lasted until 1905. The great national traumas of Sweden and Denmark in 1809 and 1814 affected the way education was discussed and eventually reformed

14. Ingmar Brohed, "Pietism och herrnhutism," in *Frihetstiden*, 69–95.

15. Parland-von Essen, "Flickors fostran"; Parland-von Essen, *Behagets betydelser*, 237–238, cf. Hammar, "Franskan," 447.

16. Parland-von Essen, *Behagets betydelser*; Öhrberg, *Vittra fruntimmer*.

17. Hammar, *Franskundervisningen i Sverige fram till 1807*, 47–66; Richardson, *Svensk utbildningshistoria*, 76; Heckscher, *Några drag*, 203–210.

18. Hammar, *Franskundervisningen i Sverige fram till 1807*, 47–68.
19. Hammar, "Franskan," 447–457; Parland-von Essen, *Behagets betydelser*, 237–238.
20. Gold, *Educating Middle Class Daughters*; Bjurman, *Catrines intressanta blekhet*; Ida Blom, "Om Pigebørns Opdragelse," *Tidsskrift for Samfunnsforskning* 20 (1979): 599–616; Gunnar Qvist, *Konsten att blifva en god flicka* (Stockholm: Liber, 1978).
21. Bjurman, *Catrines intressanta blekhet*; Possing, *Nathalie Zahle*.
22. Campe's lengthy advice-book on how to educate young girls was published in Swedish in 1798 and 1804 (two editions); it reflected the traditional Lutheran idea of the different callings of the two sexes; Bjurman, *Catrines intressanta blekhet*, 196; Ytreberg, *Nissens pikeskole*; Inger Hammar, *Emancipation och religion: Den svenska kvinnorörelsens pionjärer i debatt om kvinnans kallelse ca 1860–1900* (Stockholm: Carlsson, 1999).
23. Several of the early Scandinavian girls' schools used the term "daughters' school" in their name; it appears as a kind of generic term and was used by the herrnhuthian brother societies ("Töchterschule"), Kyle, *Svensk flickskola*, 40.
24. Kyle, *Svensk flickskola*, 40–49; Brohed, "Pietism och herrnhutism"; Arne Jarrick, *Den himmelske älskaren: Herrnhutisk väckelse, vantro och sekularisering i 1700-talets Sverige* (Stockholm: Ordfront, 1987); Hilde Sandvik, "Tidlig moderne tid i Norge: 1500–1800," in *Med kjønnsperspektiv på norsk historie*, ed. Ida Blom and Sølvi Sogner (Oslo: Cappelen Akademisk Forlag, 1999), 83–134.
25. Gold, *Educating Middle Class Daughters*; Per Nyström, *Mary Wollstonecraft's Scandinavian Journey* (Gothenburg: Acta Regiae Societatis Scientiarum et Litterarum Gothoburgensis, 1980).
26. Gold, *Educating Middle Class Daughters*, 35–37.
27. Heckscher, *Några drag*, 89–90.
28. Gold, *Educating Middle Class Daughters*, 54, 116.
29. Ibid., 82, 116, 118–121, 124.
30. Ytreberg, *Nissens pikeskole*, 18.
31. Schånberg, *Genus och utbildning*; Possing, *Nathalie Zahle*; Ytreberg, *Nissens pikeskole*.
32. Kyle, *Svensk flickskola*, 244–245, Ida Blom and Inger Elisabeth Haavet, "Familjen, staten och välfärden år 1800 respektive 1900," in *Sekelskiften och kön*, ed. Anita Göransson (Stockholm: Prisma, 2000), 17–38.
33. Kyle, *Svensk flickskola*, 40–49; Schånberg, *Genus och utbildning*.
34. Gold, *Educating Middle Class Daughters*, 60.
35. Ytreberg, *Nissens pikeskole*, 22.
36. Ibid., 23.
37. In collaboration with August von Hartmansdorff and Anders Lagergren, Kyle, *Svensk flickskola*, 44–46; Heckscher, *Några drag*, 101–104; Sjöstrand, *Pedagogikens historia*, vol. 3:2, 188.
38. Ytreberg, *Nissens pikeskole*, 38–48. The German philosophers Johann Friedrich Herbart (1776–1841) and Friedrich Eduard Beneke (1798–1854) both influenced Nissen's general educational ideas, Thuen, "Hartvig Nissen," 72–77.
39. Ytreberg, *Nissens pikeskole*, 38–48; Heckscher, *Några drag*, 101–104; Kyle, *Svensk flickskola*, 20–49.
40. Ytreberg, *Nissens pikeskole*, 38–48; Heckscher, *Några drag*, 101–106; Kyle, *Svensk flickskola*, 45–49; Blom, "Om Pigebørns Opdragelse"; Gro Hagemann, "De stummes leir?: 1800–1900," in *Med kjønnsperspektiv på norsk historie*, ed. Ida Blom and Sølvi Sogner (Oslo: Cappelen Akademisk Forlag, 1999), 135–226.
41. Ytreberg, *Nissens pikeskole*, 49–50, 61, 79–80.
42. Possing, *Nathalie Zahle*, 71–131.
43. Ibid., 137–150, 227–232, 445–446.
44. Fåhræus, *Högre Lärarinneseminariets historia*, 12–55, Karin Wilmenius, *Folkskollärarinnor i Stockholm* (Uppsala: Föreningen för svensk undervisningshistoria, 1999); Christina Florin, *Kampen om katedern* (Umeå: Acta Universitatis Umensis, 1987); Linné, *Moralen, barnet eller vetenskapen?*, 78, 97–100; Schånberg, *De dubbla budskapen*, 46–50. For Bremer, see Carina Burman, *Bremer: En biografi* (Stockholm: Bonniers, 2001), 442–457.

45. Adda Hilden, "Da kvinder lærte at lære," *Fortid og nutid* 34 (1987), 121–150; Hilden, *Fromme, stærke Kvinder.*
46. Hagemann, *Skolefolk*, 64–81.
47. Fåhræus, *Högre Lärarinneseminariets historia*, 158–193.
48. Sörensen, *Växelundervisningssällskapets normalskola;* Linné, "Morality, the Child, or Science?"; and "Female Curriculum Pioneers."
49. Ida Blom, "Nation—Class—Gender: Scandinavia at the Turn of the Century," *Scandinavian Journal of History* 21, 1 (1996): 1–16; Jill Conway, "Women Reformers and American Culture, 1870–1930," *Journal of Social History* 5, 2 (1971–1972): 164–177; Nancy Fraser, "Rethinking the Public Sphere," in *Habermas and the Public Sphere*, ed. Craig Calhoun (Cambridge, MA: MIT Press, 1992), 109–142.
50. Schånberg, *Genus och utbildning*, 59–67.
51. Richardson, *Svensk utbildningshistoria*, 68–71; Sjöstrand, *Pedagogikens historia*, vol. 3:2, 326.
52. Sjöstrand, *Pedagogikens historia*, vol. 3:2, 396.
53. Heckscher, *Några drag*, 105.
54. Possing, *Nathalie Zahle*, 457; Jytte Larsen, ed. *Alle tiders danske kvinder, Dansk kvindebiografisk leksikon*, vol. 4 (Copenhagen: Rosinante, 2000), 161.
55. Sjöstrand, *Pedagogikens historia*, vol. 3:2, 397; Ytreberg, *Nissens pikeskole*, 90.
56. Larsen, *Alle tiders*, 161, 170.
57. Ibid., 170.
58. Heckscher, *Några drag*, 203–210.
59. Schånberg, *De dubbla budskapen*, 53–63; Kyle, *Svensk flickskola*, 56, 150–178.
60. Ytreberg, *Nissens pikeskole*; Sigurd Åstrand, "Sofi Almquists samskola 1892–1936," in *Två studier av pedagogiska pionjärinsatser*, ed. Stig G. Nordström (Uppsala: Föreningen för svensk undervisningshistoria, 1985), 5–135; Florin and Johansson, *Där de härliga lagrarna gro*; Johansson, *Normalitet.*
61. Kerstin Skog-Östlin, *Att bryta ny mark* (Örebro: Örebro University, 11, 2005), http://www.diva-portal.org/oru/; Schånberg, *De dubbla budskapen*, 112–113 (accessed January 12, 2006).
62. Anne-Mette Kruse, "Pigeskole eller fællesskole?," in *Pigernes skole*, ed. Adda Hilden and Anne-Mette Kruse (Århus: KLIM, 1989), 43–73; Hagemann, *Skolefolk*, 64–87; Schånberg, *De dubbla budskapen*, 112–123.
63. Ullman, *Stiftarinnegenerationen.*
64. Ibid.
65. Åstrand, "Sofi Almquists samskola."
66. Author of the widely spread and translated book *Barnets århundrade* (Stockholm: Bonniers, 1900; New York and London: G.P. Putnam's Sons, 1909).
67. The youngest pupils even were supposed to call her "Aunt Anna," Linné, "Female Curriculum Pioneers."
68. For the public sphere, see Fraser, "Rethinking the Public Sphere"; for Scandinavia, see Ida Blom, "Refleksjoner over kjønn og stat," in *Kjønn og velferdsstat*, ed. Anne-Hilde Nagel (Bergen: Alma Mater, 1998), 24–49; and "Gender and Nation in International Comparison," in *Gendered Nations: Nationalisms and Gender Order in the Long Nineteenth Century*, ed. Ida Blom, Karen Hagemann, and Catherine Hall (Oxford: Berg, 2000), 3–26.
69. Kruse, "Pigeskole eller fællesskole?," 43; Hagemann, "De stummes leir?," 200.
70. Schånberg, *De dubbla budskapen*, 142–143.
71. Richardson, *Svensk utbildningshistoria*, 114.
72. Larsen, *Alle tiders.*
73. Kruse, "Pigeskole eller fællesskole?"; Alfred Oftedal Telhaug, *Norsk skoleutvikling etter 1945* (Oslo: Didakta, 1994); Schånberg, *De dubbla budskapen*, Johansson, *Normalitet.*
74. Ullman, *Stiftarinnegenerationen*, 210.
75. Schånberg, *Genus och utbildning*, 101.
76. Schånberg, *De dubbla budskapen*, 8.

Chapter 10

Nation-Building, Patriotism, and Women's Citizenship: Bulgaria in Southeastern Europe

Krassimira Daskalova

This chapter presents the first outline of the general development of girls' secondary education in Bulgaria from the early nineteenth century to the 1940s. It includes sections on sources and historiography, discourses on education in the Southeastern European states and on women's education in particular, and scarce empirical data. The chapter places the Bulgarian case within the Southeastern European context and illustrates how nation-building intersected with the development of girls' education.

The modern history of the Balkan region began with the formation of several small nation-states: Serbia (1815), Greece (1830), Romania (1859), and Bulgaria (1878), and later, Albania (1912), which emerged through secession from the Ottoman Empire. The borders changed several times due to the so called Eastern question about the legacy of the Ottoman Empire. In these countries there was a strong stress on education as part of nation-building and modernization (Europeanization, Westernization), both before and after the establishment of the states. Within limits, this stress also extended to women.

Although sources exist (including official and private archival collections, memoirs, diaries, newspaper materials, brochures), girls' secondary education in the Ottoman Empire and its successor nation-states in Southeastern Europe has not yet been studied systematically.[1] This stems in part from the uneven development of social and cultural history in Central, Eastern, and Southeastern Europe and of women's and gender history in particular. Girls' secondary education in the Balkans is briefly discussed in more general historical works, including Nicolae Iorga's history of Romanian education and Roumen Daskalov's two volumes on the modernization of Bulgarian society, 1878–1939.[2] Some entries in a new biographical dictionary dealing with women's movements and feminisms in Central, Eastern, and

Southeastern Europe contain information about the development of girls' education throughout this region.[3] Material can also be gleaned from doctoral theses, books, chapters, and articles, some of which are written from a feminist perspective.[4]

Sidiroula Ziogou-Karasterghious focuses on the education of Greek women during the nineteenth century, as do dissertations by Eleni Fournaraki[5] and Stéphanos Kavallierakis.[6] Women's and gendered education feature in the volume on gender and education in Greece edited by Vasiliki Deliyanni and Sidiroula Ziougou.[7] Katerina Dalakoura has studied the education of women from Greek ethnic communities within the Ottoman Empire (i.e., those outside the Greek nation-state, established in 1830) from the nineteenth century to 1922.[8]

Elif Ekin Akşit explores the paradoxical connections between girls' education and nationalism in the late Ottoman Empire and the early Turkish republic.[9] Few publications focus on girls' education in Serbia, but information can be found in monographs and collections dealing with the relationship between the modernization of Serbian society, the "woman question" and women's situation as a measure of modernization.[10] Similarly, a few texts touch on women's/girls' education in Romania, notably the collection of documents edited by Stefania Mihăilescu.[11] Alin Ciupală's chapter on ideals and realities discusses women's education in nineteenth-century Romania,[12] while Traşcă's case study of the lyceum "Principesa Ileana" in Cluj-Napoca deals with the period 1926 to 1948.[13]

I contributed materials about girls' education in Bulgaria to a documentary volume on women in modern Bulgarian society and culture and have published on discourses of Bulgarian education, and Bulgarian teachers, before and after 1878.[14] The role of Catholic nuns in the development of girls' education in the Bulgarian town of Rousse is the subject of a recently published pamphlet.[15] Three periods can be identified in the education of Bulgarian girls: first, the so-called National Revival prior to the liberation from the Ottoman domination and the establishment of the Bulgarian nation-state in 1878; second, the bourgeois nation-state between 1878 and 1944; and third, state socialism from 1944, when the communists took over, until 1989. In this chapter, discussion focuses on the first two periods, as research remains to be done on the period from 1944.

Early Nineteenth-Century Debates about Women's Education in the Balkans

Nineteenth-century developments in Bulgaria were strongly influenced by Greek, Serbian, Russian, West European (mostly French), and American educational ideas and institutions. Up to the 1840s, the most important influence was Greek, not only in the types of schools and programs but also in the number of people (mostly men) educated in Greek institutions of elementary and higher learning. In his "Constitution" (1797), the Greek national revolutionary and poet Rigas Velestinlis (1757–1798) supported the idea that schools for both boys and girls should be established in each village. However, it was only after the Greek revolution of 1821 and the subsequent independence and establishment of the Greek state that specific (though Western Enlightenment-inspired) discourse on women's education was born.[16]

Discourses on women's education in nineteenth-century Greece were part of a wider nationalist ideology aimed at promoting modernization. Within Greece two attitudes coexisted: one that accepted coeducation in elementary schools, and a second that insisted on the establishment of separate schools for girls (both public and private). While wide differences existed between the Greek regions, the most visible were those between the culturally conservative, traditional, continental part of Greece and the more developed and urbanized ports and administrative centers. As elsewhere, the rural population considered girls' education absurd and only the urban population was receptive to the model of separate girls' education. In the later 1820s the private initiative of Greek women, American Protestant missionaries, and certain municipalities supported a few Bell-Lancaster (so-called mutual) elementary schools. In the 1830s the first "hellenic" (private, secondary) schools for girls were established in Hermoupolis (Cyclades) and in Athens.[17] In 1834 the Greek government introduced compulsory girls' and boys' primary education from ages five to twelve years. The law of 1834 defended separate education for girls and boys without banning mixed elementary schools; however, in 1852, the Greek government forbade coeducation in primary schools.[18]

Despite this legislation, the Greek State left girls' secondary education in private hands and invested energy and resources in the education of boys. Private schools reserved for girls from well-to-do families existed in urban regions. In contrast, boys' secondary education was free. Opposition to these institutions came from conservatives and modernists alike. Conservatives saw women's education as a principal agent of the Westernization of morals and manners and thought women's education and immorality were synonymous. Greek modernists supported Westernization but tended to reject Westernized girls' education in the private secondary schools.[19] At the same time, the idea of woman's mission or vocation as mother-educators in spreading national ideology reinforced views in support of a different kind of women's education. In Athens, Hermoupolis, Patras, and other urban centers during the nineteenth century, several "superior" girls' private schools were established. Particularly popular were the *Arsakeions*, established in Athens, Corfu, and Patras and run by the "Society of Friends of Education." Named after Apostolos Arsakis, a generous donor, the *Arsakeions* offered formal secondary education to girls, trained future women teachers, and contributed to women's sense of autonomy. During the second half of the century, Western missionary educational institutions opened for Greek girls. The sisters of Saint-Joseph de l'Apparition set up Catholic schools in Greece in 1856.[20] This presence of foreign Protestant and Catholic schools alongside secular and national institutions created a range of opportunities for girls in an area where Orthodoxy and Islam played a major role in local cultural traditions.

Men and women did not enjoy the same rights as citizens in the Balkans. Nineteenth-century educators from the region, like their Western counterparts, believed that women and men were essentially distinct creatures, possessing different bodies, which implied different behavior, abilities, and intellectual potential. They insisted on a school curriculum for girls that would correspond to women's "natural" domestic and motherly duties.

During the 1870s, groups such as the Association in Favor of Women's Education, founded in Constantinople in 1871, and the Ladies Association in Favor of Women's Education, established in Athens in 1872, as well as wealthy

Greeks from the diaspora, played a role in founding several new schools. These were based on the idea of a specific woman's "nature," "role," and "vocation," and perpetuated a belief in specific education for girls different from that for boys. The ideal of the mother-pedagogue encouraged the development of teaching as a woman's profession, viewed as a mission to serve the nation. Politicians, intellectuals, and women pedagogues, including Sapho Leondias (1832–1900) and Kalliopi Kechayia (1839–1905), supported this discourse. Other educated Greek women expressed concerns about the social mission of "the woman" and discussed women's important role in the "revival" of the Greek nation. They published articles and pedagogical works, gave public lectures, and edited the first women's periodicals that reflected on the vocation and education of the female sex.[21]

These initiatives should be understood in relation to two negative stereotypes that persisted in Greece during the nineteenth century: the illiterate backward woman who constituted the main obstacle to the progress of the nation and symbolized Greece's failed attempt to become "European"; and the *mondaine* graduate of the *parthenagogion* (a general term referring to girls' schools of all levels) who was frivolous and vain.[22] The latter possessed useless knowledge, parroted foreign modes, was shallow, proved highly unreliable as a mother, and possessed dubious sexual morality. In tune with the patriarchal and nationalist spirit of the time, the prevailing mentality believed girls should be educated to become "mothers of the nation" while boys' education should produce "citizens."[23] The first public secondary schools for girls, equivalent to those for boys, were opened in 1917, but it was only in 1929 with the educational reform of Eleutherios Venizelos (1864–1936) that secondary education for girls became comparable with that of boys.[24] The late development of girls' secondary education was in line with expectations and ideas about women's role and place within Greek society.

In neighboring Serbia, the enlightened thinker Docitej Obradovic (1742–1811) was the first to write about women's emancipation. He advocated coeducation and saw overcoming women's ignorance and "barbarism" as an important step in Serbian development. During the 1840s the first elementary schools for girls were opened in the capital city of Belgrade, and subsequently in most territories of what would in the twentieth century become Yugoslavia. In 1853 a private Educational Institute for Girls opened in Belgrade with the aim of training Serbian female teachers and preparing women to become responsible nurturers of the future citizens of the Serbian fatherland. The first modern women's high school (*Visa zenska skola*) in the Balkans was established in the Serbian capital in 1863.[25] Several Bulgarian girls received their secondary education at this school, where they also trained as teachers, nationalists, and, quite paradoxically, as patriots. Both Bulgarian boys and girls who studied in Greek and Serbian schools learned patriotism there and transmitted the "national idea" to the Bulgarians. These included some of the first well-educated Bulgarian teachers: Zyumbula Talimova, Zhelka Ivanova, and Smarayda Nacheva from Eski Zagra (today Stara Zagora). Bulgarian-Serbian educational cooperation was short-lived, however, due to the escalation of both Serbian and Bulgarian nationalisms in the 1860s and 1870s.

Nationalism and Women's Education in Bulgaria up to 1878: A Patriotic Duty to the Nation

During the period of National Revival, education was entirely a matter of self-organization and owed a great deal to the initiative of local communities. These exercised a measure of control over education and teachers through the institutions of local self-government (village or town councils, the so-called *obshtina*) and through special school boards (*uchilishtni nastoiatelstva*), composed of local notables and representatives of the guilds.

As in the other regions of Southeastern Europe, the idea of women's education asserted itself slowly and against the resistance of the traditional patriarchal values in Bulgarian society under Ottoman domination. School teaching was the only profession open to educated women. From the 1840s until the establishment of the nation-state in 1878, the periodical press was largely published by Bulgarian intellectuals in the émigré colonies outside ethnic Bulgarian territories. It circulated emancipatory ideologies, including nationalism, liberalism, socialism, and feminism, and avidly discussed "the woman question," and particularly the necessity of women's education.

Similar to their counterparts in other European countries (especially in France), Bulgarian male writers and national activists supported the idea of women's education and argued that women should be educated to become better housewives and mothers, as well as pleasant and entertaining companions for their husbands. Secondly, education had a patriotic goal, as women should mobilize around national goals and raise their children as future citizens of the nation. Drawing on Rousseau's ideas, the writer and national activitist Petko Slaveikov argued that women should be educated to serve individual men and "become a man's good comrade." Slaveikov also insisted that as "the man, the family and the whole nation depend on the instruction and the education of women," women's education should be supported even more than that of men.[26] Parallel with this discourse on women's education, translations of normative and moral works, mostly French and Russian but also American, appeared together with original texts by literary men. These aimed at cultivating the taste for "good" reading among young girls.[27]

Lyuben Karavelov (1834–1879), a Bulgarian national revolutionary and man of letters, was one of the few intellectuals in the Southeast European context who conceived of women's education in terms of natural rights. He was fiercely critical of special training for girls, which he argued incapacitated their minds and killed their independence. Karavelov attacked the European (mainly the French) educational system that he thought, in contrast to the American, made women "trained slaves," "beautiful dolls," to be used by "old children" and "whiskered masters."[28]

The movement for women's emancipation among Bulgarians began in the 1840s with the opening, under Greek influence, of several dozen secular elementary girls' schools run along "mutual" or Bell-Lancaster lines. These were supported financially by local women's initiatives and some rich Bulgarian émigré merchants. Anastasia Dimitrova established the first school of this type at Plevna (today's Pleven) in 1841. By the beginning of 1850, thirty-five schools were in existence, rising to ninety by 1878. In addition, twenty-nine Catholic and several Protestant day schools were opened after the Crimean War of 1853–1856. Although the Catholic schools were considerably more numerous than the Protestant schools, the Bulgarians considered the latter to be superior and elites preferred to send their boys and girls to them. More Protestant schools were founded subsequently, including some high schools, which continued to function in the twentieth century.

Some women teachers openly advocated women's "uplifting" and progress through reading and education. In 1853 Stanka Nikolitsa-Spaso-Elenina translated texts of the Serbian Docitej Obradovic into Bulgarian. Rada Kirkovic attended the *Fundukleevskaja gymnasia*, in Kiev, Russia, which was well-known among the South Slavs. She left with a gold medal and later translated geographical textbooks for the Bulgarian secondary schools. Yordanka Filaretova, the president of one of the most active Bulgarian women's societies in the 1870s–1890s, was a strong supporter of women's education. From the late 1850s, the first women's benevolent, philan-thropic and educational organizations opened women's reading clubs and libraries, organized public lectures, sent articles to periodicals, and participated in amateur theaters. They also gave stipends for the education of the most promising schoolgirls and sent some of them abroad to Slavic schools, most often in Serbia and Russia, but also in Czech-speaking educational institutions.

Upon their return to Bulgaria all these women taught and opened some of the best-known girls' *klasni* (*class* schools). *Class* schools had one to five classes above ele-mentary schooling and were the prototype of full secondary education. Within girls' *class* schools the curricula combined more general courses, such as Bulgarian gram-mar, church history, Old and New Testament, geography, mathematics, ancient history, world history, natural history, physics, French language and pedagogy, with handwork and needlework, which was considered to be of the highest importance for future housewives and mothers.[29]

By 1878 there were nineteen *class* schools opened mostly by young women who had undertaken secondary education in Serbia or Russia. Most of the first women teachers educated abroad were daughters, sisters, or wives of teachers and came from intellectual families.[30] These included Talimova, Ivanova, and Nacheva-Kolarova, who were educated in Belgrade with the support of the Serbian government. During the 1860s and 1870s an important part of the future women's intelligentsia was edu-cated in Russia, including Anastasia Tosheva, Sanda Milkova, Tsarevna Miladinova-Alexieva, Rada Kirkovic, and Ekaterina Karavelova.[31] They left diaries and memoirs charting their studying- and teaching experience.

There were fewer Bulgarian *class* schools for girls than for boys due to the rural character of Bulgarian society and traditional patriarchal hostility to girls' educa-tion. Given this fact, some liberal parents allowed their daughters to attend boys' *class* schools, either within the male classes or in special classes for girls, as occurred in the town of Shoumen in 1874–1875. While male teachers were invited to teach at *class* schools for girls, the reverse was unthinkable, even when women had attained a similar level of education.

The Bulgarian *class* school for girls established by Anastasia Tosheva in Eski Zagra in 1863 shows the interplay between nationalism, modernization, and the emancipation of women. Tosheva was one of the few Bulgarian girls educated in

Odessa, thanks to Russian state funds. She opened her *class* school only six months after the American Protestant missionaries, Theodore Byington and his wife Margaret, a college graduate,[32] founded the mission school in Eski Zagra. The latter, set up on the American Mount Holyoke model, aimed to educate girls, train Bulgarian women teachers, and inspire them to support the missionaries' plan of religious and educational reform. The curriculum included reading, writing, arithmetic, grammar, geography, vocal music, needlework, and daily instruction in the Bible.[33] Although the missionary school was a real success in academic terms and helped develop girls' education, no students converted to Protestantism and the Byingtons made no progress in evangelizing in the town. As Tosheva wrote in her autobiography, she founded her *class* school in direct response to the Byingtons' initiative, within the context of rising Bulgarian national sentiments. Members of the local governing body, the *obshtina*, saw this mounting Protestant presence with concern and encouraged Tosheva to open a Bulgarian girls' school in Eski Zagra to "oppose the foes of Orthodoxy and our nation."[34] In her opening speech Tosheva insisted that if Bulgarians wanted progress and prosperity they had to educate their daughters. She was one of the first women radical supporters of the idea of women's education as a path to a successful future for the Bulgarian nation, and her speech heralded the huge public debate on "the woman question" that developed in the following decades. During the academic year 1868–1869, Tosheva's *class* school became the first Bulgarian women's school where students completed a five-year course of study.

The girls' *class* school teachers in the towns of Gabrovo and Plovdiv followed Tosheva's example in the following years, and opened five classes in their girls' schools. These three *class* schools gave the most advanced education for women in Bulgaria prior to 1878. Most girls' *class* schools, however, offered only two, three, or four years of education. Full secondary schools for women, called *gymnasii* (with six classes above elementary education), opened after 1878.

Teachers within the Bulgarian Revivalist Intelligentsia (Until 1878)

The only profession open to women under Ottoman rule was that of teacher. There were about 400 women teachers out of a total of 4,378 teachers prior to 1878. Together with the clergy (3,623 persons during the "Revival"), teachers formed the bulk of the Bulgarian intelligentsia, which comprised about 9,000 persons in total.[35] Women teachers on average had fewer opportunities for serious study than men.

The most highly educated were the graduates of secondary schools. Twenty finished Russian schools, six completed Serbian schools, and five Czech schools. In addition, seventy-seven women had attended Bulgarian *class* schools. The number of teachers who graduated in Russia was the consequence of deliberate Russian policies of giving grants to Southern Slavs in order to acquire cultural and political influence, although the Russian government gave much more support for male than female students from the Balkans. The Russian graduates then propagated new pedagogical ideas and "progressive" methods in Bulgaria, took part in educational reform, and helped establish *class* schools and *gymnasii*. Czech and French institutions played a role in the preparation of future Bulgarian women teachers, and a few Bulgarian women teachers from the pre-1878 period studied in Romanian schools. No Bulgarian teachers studied in Ottoman schools due to the Islamic character of the educational institutions and the growing patriotic spirit against Ottoman rule, although most of the Bulgarian intelligentsia could speak (and often write) Turkish.

Foreign men and women teachers also taught in Bulgarian schools[36] and in missionary and private urban schools, before and after the Liberation in 1878. Some Serbian and Czech women advocated Slavic unity and progress. After the Crimean War, several Catholic and Protestant schools were established in the Ottoman Balkans, including in Bulgarian lands. Most teachers in the Protestant schools were Americans. The presence of foreigners among the pedagogues in the Bulgarian schools demonstrates a rich mix of teachers who promoted various "progressive" educational ideas that were little known at the time in the local contexts.

A rudimentary professional and ethical "code" for (both male and female) teachers began to develop over the course of the nineteenth century shaping an idealized image of the professional educator. Increasing professionalization was illustrated by the establishment of a formal framework for the work of teachers (including written employment contracts and regulation of the working day and school-year); the adoption of "progressive" teaching methods and new curricula; changes in the organization of schooling, management, and discipline; and the establishment of professional associations.

Teachers acquired a public role during the era of the National Revival, with its twofold task of modernization and formation of a national consciousness. In this context, teaching came to be perceived as a missionary activity, a "calling," and an activity of "Enlightenment" (in Bulgarian *prosveta*, *probuda*) for the spread of knowledge among the uneducated. However, for the missionaries of the National Revival, Enlightenment meant primarily Enlightenment *to* a national consciousness. Teaching was to arouse national pride and a sense of belonging to a nationwide ("imagined") community. The formation of national consciousness was ascribed especially to women, first as wives and housewives and then as mothers and educators of the future citizens of the Bulgarian nation. In accordance with the patriotic function prescribed to education, knowledge of the Bulgarian language and Bulgarian (medieval) history was of particular importance. Teaching could thus be perceived as the patriotic task of rousing in children love for the Fatherland.

Girls' Education in Bulgaria after 1878

The Bulgarian nation-state was established in 1878 and Bulgarian nationalism institutionalized gender difference based on gender hierarchy. The first Bulgarian *Turnovo* Constitution of 1879 introduced universal schooling and compulsory elementary education for all Bulgarian citizens. Following West European examples, the State accepted the principle of separation in the education of girls and boys. The primary school curriculum copied some elements of the French, Belgian, Swedish,

Russian, Austrian, and German school systems but adapted them to the "sober Bulgarian tradition."[37] Coeducation was allowed only if girls were less than eleven years and boys less than twelve years old. The State opened several new middle and secondary schools for girls and boys after 1878 and the number of girls who progressed through elementary schooling rose.

Secondary education in the Bulgarian nation-state replicated the Napoleonic system. It favored boys' secondary education and paid less attention to the girls' educational needs. From the elementary school to the high school, girls and boys studied different subjects that were believed to be "more suitable" for their "nature" and their social roles. Thus Bulgarian (male) educational authorities strengthened the view that men were inherently political and women naturally domestic. Coeducation was not permitted in the secondary schools, although some pedagogues supported the idea and in the early 1890s published journals (such as *Pedagogium*, 1891–1894) that promoted this policy.[38] The existing pedagogical periodicals propagated the history of various kinds of educational ideas—from the classical texts of J.A. Komenski, John Locke, Jean-Jacques Rousseau, Johann Heinrich Pestalozzi, Friedrich Fröbel, and Herbert Spencer via Herbart's pedagogy and anti-Herbartian criticisms, to Ellen Key's new education texts.

The statistical data on Bulgarian secondary education is complicated until 1909 and difficult to compare. There were various kinds of schools with a different number of "classes," called, as in the pre-1878 period, *klasni* (*class*) schools. There were also incomplete and complete types of *class* schools, called *gymnasii* (modeled after the German *Gymnasien*). The schools were both single-sex and coeducational. Most schools were called *narodni* (people's) schools and were supported by the Bulgarian state. There were also private schools not supported by the State, for example, for ethnic minorities and religious groups. In 1885–1886 there were seven complete *gymnasii* for boys and two complete *gymnasii* for girls. But women's *gymnasii* had six classes while men's had seven classes and different curricula.

It was only in 1891 that the so-called Zhivkov's law (named after Minister Georgi Zhivkov) signaled the start of centralization by the State of both primary and secondary schools under the ministry of education. The government unified the curricula and textbooks for the various kinds of schools. Teachers were now deprived of the freedom to choose what and how to teach and became increasingly powerless state employees in a large bureaucratic machine, subject to regulation, supervision, and decisions from above. There was a decrease in teachers' incomes and a general loss of prestige for the profession. In addition, salaries of village teachers were 10 percent lower than those of urban teachers, and a trend toward feminization of the profession gradually set in, consistent with its subordinate status. Women teachers received 10 percent less in salary than male colleagues.[39] An administrative decree in 1899 prohibited married women from teaching. While this was abrogated in 1904, rising unemployment in 1927 saw a law enacted that dismissed married women teachers at the age of forty. Socialism and other ideologies replaced nationalism for some, while indifference and depoliticization gained others. Teaching no longer represented a mission, just a living. Despite the activity of the Bulgarian Teachers' Union, which emerged in 1895 to defend both, the professional and political rights of teachers as citizens, the teachers' public role deteriorated.

The Zhivkov law reconfirmed gender hierarchies in secondary education, with boys' schools containing seven classes above the primary level and girls' schools six classes and a different curriculum. This deprived girls of access to national university education. Immediately after the passage of a new University law in parliament in 1895, several hundred Bulgarian women (most of them teachers) sent petitions to Parliament and applications to *Vissheto uchilishte* (the Higher school) in Sofia, which had been established in 1888 and would be renamed a University in 1904. They pleaded that it should be opened to women. Women's petitions were neglected on the grounds that girls were not sufficiently prepared to enter the highest educational institution in Bulgaria; according to state officials, their high schools did not provide a full secondary education. Women made a new collective appeal to the National Assembly for the equalization of girls' high schools with boys' and for women's admission to the university. They sent an open letter to all the daily newspapers in Sofia and provoked a heated public debate on women and university education. A large number of polemical articles on women's education were published. One particularly critical voice came from the Bulgarian feminist of Ukrainian origin Lidiya Shishmanova (1865–1937), married to one of the best-known Bulgarian intellectuals, Prof. Ivan Shishmanov. She showed that conservative opinion was untenable and concluded that sooner or later Bulgarian women would enter the university, like women in many other countries.

The first outcome of the campaign was the 1897 law, which equalized secondary school education for girls and boys, with seven grades for all. Girls' secondary schools now comprised a five-year lower section and a two-year upper section divided into two subsections, pedagogical and general. The aim of the former was to prepare women teachers for the middle schools, while the latter provided the necessary knowledge for girls who wanted to continue their education.

The 1904 Law for Secondary and Girls' Education underpinned the opening of seven women's *pedagogicheski* (pedagogical) schools equal to those for men. In tune with dominant ideas about women's and men's qualities and so-called natures, girls were considered the most appropriate to become teachers in kindergartens. A special course for training kindergarten teachers was opened within the girls' pedagogical school in Sofia in 1905. In 1912 the Bulgarian state also opened a Turkish pedagogical school with the special aim of preparing Muslim teachers for the elementary schools of the Turkish minority.

In 1909 Moushanov's law (named after Minister of Education, Nikola Moushanov) reconfirmed the compulsory character of primary schooling, which was extended to fourteen years of age. The law introduced a new level of education with the opening of the so-called *progymnasii*, which bridged the elementary school level and *gymnasii*. The *progymnasii* were free, but still not compulsory. They gave general education and avoided the early professional orientation of the students. Male and female teachers for *progymnasii* were prepared in special two-year higher pedagogical courses, which were opened within several boys' *gymnasii*.

The complete secondary school level included institutions that provided general education: *gymnasii* (real, classical, and semiclassical, i.e., Latin only), and pedagogical and technical schools. The pedagogical schools combined general education with special training, while technical schools prepared personnel for industry and agriculture. This law, like the previous ones, stressed the primary role of general education, which suited best the necessities of the egalitarian Bulgarian society and gave the sons and daughters of the lower social strata in the countryside an opportunity to climb the social pyramid.[40]

According to the 1909 law, each Bulgarian province could have only two *gymnasii*, one for girls and one for boys. While the greatest part of the curriculum was the same,[41] girls did not study Latin, Greek, descriptive geometry, art, and political economy. In their place was more home economics and handwork, subjects that were considered suitable for their "natural" predispositions (see table 10.1).

Progymnasii and *gymnasii* increased in the first decade of the twentieth century. However, the number of girls studying in the existing *gymnasii* was half that of boys. While coeducational schools spread, mixed complete *gymnasii* did not exist.

Bulgarian schooling headed in a new direction in 1921 with a law named after the Minister of Education, Omarchevski. Within the post-First World War economic and ideological crisis, the short-lived rule of the agrarian party (1920–1923) aimed to redirect the State's efforts from general to professional education, in line with what they believed to be practical necessities of the Bulgarian economy and society.[42] Supporters of professional education argued that general education prepared young people inappropriately and created an "intellectual proletariat" who

Table 10.1 Curricula of men's and women's *gymnasii*, according to the 1909 law

Subject	Boys	Girls
Bulgarian language and literature	√	√
Old Bulgarian language and literature	√	√
Philosophical propedeutics	√	√
Russian, French, or German languages (by choice)	√	√
Latin	√	
Greek	√	
Mathematics and descriptive geometry	√	
Mathematics		√
Bulgarian and world history	√	√
Geography	√	√
Political economy	√	
Physics	√	√
Chemistry	√	√
Natural history	√	√
Hygiene	√	√
Art	√	
Home economics and handwork		√
Music	√	√
Gymnastics	√	√
Stenography (optional)	√	√

Source: *Istoria na obrazovanieto I pedagogicheskata misul v Bulgária*. vol. 2 (Sofia: Narodna prosveta, 1982).

looked for state employment only. The Omarchevski law envisioned the opening of a series of lower and higher special and professional schools affiliated to various handicrafts and industries. This reform extended compulsory education for both sexes from four to seven years, thus from age seven to fourteen. The *progymnasii* became the mass type of (compulsory) educational institution, and led to a raise in educational levels. The *realka* was introduced to bridge elementary schools and *gymnasii*. It provided a more economical secondary education for peasant children (four years elementary education, three years *progymnasia*, followed by an optional three years' professional education, whereas the *gymnasia* already encompassed five years after the *progymnasia*). The number of *gymnasii* was cut, but the number of state *progymnasii* almost tripled between 1919 and 1927.[43] In 1920, girls formed only 25 percent of the students in this type of school, rising to about 33 percent in the late 1920s. A tendency toward gender inequality was supported by the smaller number of girls' *gymnasii* and their lower enrolments.[44]

The lower professional schools followed after the *progymnasii* with two years of education. New specialist secondary schools in various fields were established for girls and boys who had completed the *realka*.[45] The first girls' vocational school was opened by the women's society *Maika* (Mother) in Sofia in 1893 and offered a three-year education in dressmaking and cookery. It merged in 1909 with the existing state school for millinery, horticulture, and embroidery, adding teacher training from the school year 1910–1911. The Red Cross opened a professional training school for nurses in Sofia in 1895 that continued to exist through the 1940s. The number of agricultural schools for girls grew from one in the late nineteenth century (compared with four for boys) to thirteen in 1940 (compared with twelve for boys). One new general trade school for girls (and one for boys) was established in Sofia in 1926–1927, in addition to the seventy-nine girls' practical schools for housekeeping and handicrafts (compared with seventy-four boys' handicraft schools) that existed in interwar Bulgaria, and the coeducational secondary level state music school established in 1912 that developed into the State Musical Academy by 1922.

A number of private schools not supported by the Bulgarian State were also established. The first private commercial school for girls (later renamed commercial *gymnasia*) was opened in 1916 and existed until 1926–1927. Its creator, Anna Karima (1871–1949) was the first president of the Bulgarian Women's Union. According to the available state statistics (which are not consistent in recording the gender profile of schools, students, and teachers), most of the private schools were at elementary level, but in 1919–1920 there were also thirty-six private *progymnasii* and eight private *gymnasii*; in 1938–1939 the number of private *progymnasii* and *gymnasii* was fifty-six and twenty, respectively.[46] The most numerous among them were the schools of the various minority groups (including Turkish, Greek, Jewish, and Armenian) and the missionary schools attended by children of Bulgarian citizens (including Catholic and Protestant: thirty-one French, twenty-two German, seven American, and Italian). German education in Bulgaria proliferated especially during the interwar period due to the widening economic, political, and cultural contacts between the two countries; twice as many Bulgarian students attended German schools than attended French schools.[47]

The lack of financial resources in many towns and larger villages after 1878 meant that girls who wanted higher education were allowed to attend boys' schools. Especially after the turn of the twentieth century when the first Bulgarian women were admitted as regular students to Sofia University in 1901, and after the *Moushanov* law (1909) introduced *progymnasii* level of education, the Bulgarian educational authorities were more open to coeducation. They wanted to see mixed *progymnasii* and *gymnasii*, as the coeducational principle had already been successfully introduced into the elementary schools and at the University. Particularly at the *progymnasii* level, coeducation was brought into existence in the face of the practical needs of various localities as it involved a more frugal use of resources. It also gave the opportunity for both boys and girls to attend the closest school in their neighborhood. At the start of 1920, the Bulgarian Ministry of education instructed all state *progymnasii* to become coeducational institutions.[48]

Evaluating the achievements of education in the process of Bulgarian modernization shows substantial progress in the education of both girls and boys and in general literacy rates. As tables 10.2 and 10.3 demonstrate, however, there remained differences up to 1946 in literacy rates between men and women, between urban and rural populations (as was characteristic of all traditional peasant societies), and between Bulgarians and ethnic minorities, especially the Turkish and Gypsy minorities.

Increasing education and rising literacy rates were due, amongst other factors, to the advancing integration of Bulgaria into the European cultural arena and to the struggles of women activists and male supporters of women's emancipation. It was also due to the activities and criticisms of women activists and feminist

Table 10.2   Male and female literacy rates, 1900–1946 (%)

Year	General	Men	Women
1900	29.5	45.0	14.0
1920	52.8	66.4	39.2
1934	68.4	79.6	57.2
1946	75.6	83.9	67.3

Source: Krassimira Daskalova, "Gramotnost," in *Bulgarska kniga. Encyclopaedia*, ed. Ani Gergova (Sofia-Moskva: PENSOFT, 2004), 130–132.

Table 10.3   Urban and rural literacy rates, men and women, 1900–1946 (%)

Year	Towns			Villages		
	Men	Women	General	Men	Women	General
1900	67.3	39.7	53.5	38.9	7.4	23.15
1934	88.1	73.7	80.9	77.1	52.5	64.8
1946			85.5			72.3

Source: Krassimira Daskalova, "Gramotnost," in *Bulgarska kniga. Encyclopaedia*, ed. Ani Gergova (Sofia-Moskva: PENSOFT, 2004), 130–132.

organizations in the country. Their first step was securing opportunities for better education for girls.

Girls' Education in Bulgaria after 1944

The establishment of the communist regime in Bulgaria with a dominant ideology of gender equality led to the literacy rate for both men and women improving rapidly. The widening of the school system, with its free primary and secondary education, resulted in a higher percentage of girls being enrolled in institutions of secondary and university education. It was a step toward women's greater autonomy and emancipation. The fact that all this was done within a totalitarian system complicates the issues and points to the need for a separate and ideally a comparative East European project.

Notes

1. In this text, secondary education generally encompasses the period between the fifth and twelfth year of education. Although class is not absent for the period under discussion, especially in the case of Bulgaria in the egalitarian peasant society prior to the Liberation in 1878, all secondary schools were open to children of both poor and well-off families. After 1878 the nation-state paid more attention to the education of urban populations and neglected the interests of peasant children until the early 1920s when the government of the peasant leader Alexander Stamboliiski tried to change this situation.
2. Nicolae Iorga, *Histoire de l'enseignement en pays roumains*, trans.Alexandrine Dumitrescu (Bucharest: Edition de la caisse des ecoles, 1932); Roumen Daskalov, *Bulgarskoto obshtestvo, 1878–1939.* vol. 2: *Naselenie. Obshtestvo. Kultura* (Sofia: Guttenberg, 2005).
3. Francisca De Haan, Krassimira Daskalova, and Anna Loutfi, eds. *A Biographical Dictionary of Women's Movements and Feminisms, Central, Eastern and South Eastern Europe, 19th and 20th Centuries* (Budapest and New York: CEU Press, 2006).
4. I am very grateful to Asli Davas, Elif Ekin Akşit, Eleni Fournaraki, Katerina Dalakoura, Alexandra Bakalaki, Roxana Cheşchebec, Dubravka Stojanović, Ivana Pantelić, Rebecca Rogers, and Joyce Goodman for providing me with bibliographical data.
5. Eleni Fournaraki, "Institutrice, femme et mère": idées sur l'éducation des femmes grecques au XIXe siecle (1830–1880)," vol. 1–2. Ph.D.diss. University of Paris V11 1992.
6. Stéphanos Kavallierakis, "Éducation et écoles étrangères en Grèce au XIXe siècle. Le cas des écoles des sœurs de St-Joseph de l'Apparition 1856–1893," vol. 1–2. Ph.D.diss. University Marc Bloch–Strasbourg 2008.
7. Vasiliki Deliyanni and Sidiroula Ziougou- Karasterghiou, eds. *Ekpaideysi kai Fylo* (Thessaliniki: Vanias Press, 1993), 71–129, 193–209, 303–417, 417–445; see also Sidiroula Ziogou-Karasterghiou, ' "Demoiselles sages et mères parfaites": Les objectifs des écoles pour filles et la politique de l'enseignement au 19ème siècle,' in *Historicité de l'enfance et de la jeunesse, Actes du colloque international* (Athens: Archives historiques de la jeunesse grecque, 1986), 399–416.
8. Katerina Dalakoura, *H Ekpaidefsi ton Gynaikon stis Ellikes Koinotites tis Othomanikis Aftokratorias (19os ai.—1922): koinonikopoiisi sta protypa tis patriarchias kai tou ethnikismou 2* (Athens: Gutenberg Press, 2008).
9. Elif Ekin Akşit, "Girls' Education and the Paradoxes of Modernity and Nationalism in the Late Ottoman Empire and the Early Turkish Republic," Ph.D.diss, State University of New York at Binghamton, 2004; Elif Ekin Akşit, *Kızların Sessizliği. Kız Enstitülerinin Uzun Tarihi* (İstanbul: İletişim Yayınları, 2005).

10. Neda Bozinović, *Zensko pitanje u Srbiji u 19. i 20. veku* (Belgrade: Devedesetèetvrta, 1996); Latinka Perović, ed. *Srbija u modernizacijskim procesima 19. i 20. veka, 2, Položaj žene kao merilo modernizacije* (Belgrade: Institut za noviju istoirju Srbije, 1998). This book by Perović includes three texts (by Perović, Trnavac, and Nikolova) dealing with women's education, plus summaries in English: Latinka Perović, "Modernost i patrijarhalnost kroz prizmu drzavnih zenskih institucija: Visa zenska skola (1863–1913)"; Nedeljko Trnavac, "Inferiornost prema skolovanju zenske mladezi u Srbiji do 1914"; and Maja Nikolova, "Skolovanje zenske mladezi u Srbiji do 1914"; Latinka Perović, ed. *Žene i deca: Srbija u modernizacijskim procesima 19. i 20. veka* (Belgrade: Helsinški odbor za Ljudska prava u Srbiji, 2006); Ljubinka Trgovcević, *Planirana elita. O studentima iz Srbije na evropskim univerzitetima u 19. veku* (Belgrade: Istorijski institut, 2003).
11. Stefania Mihăilescu, ed. *Emanciparea femeii române. Antologie de texte.* vol. I (1815–1918) (Bucharest: Editura Ecumenică, 2001).
12. Alin Ciupală, *Femeia în societatea românească a secolului al XIX-lea: între public și privat* (Bucharest: Editura Meridiane, 2003).
13. O. Trașcă, "Aspecte ale educației femeii în România în perioada 1926–1948. Studiu de caz: Liceul de fete 'Principesa Ileana' din Cluj-Napoca," in *Prezente feminine. Studii despre femei in Romania*, ed. Ghizela Cosma, Eniko Magyari Vincze, and Ovidiu Pecican (Cluj: Editura Fundatiei Desire, 2002), 101–134.
14. *Istoria na obrazovanieto I pedagogicheskata misul v Bulgaria.* vols. 1–2 (Sofia: Narodna prosveta, 1975 and 1982); Krassimira Daskalova, "Obrazovanie na zhenite I zhenite v obrazovanieto na vuzrozhdenska Bulgaria," in *Godishnik na Sofiiskia Universitet* (Sv. Kliment Ohridski," Centur po kulturoznanie 1992), 85, 5–18; Daskalova *Literacy and Reading in 19th Century Bulgaria*, in The Donald W. Treadgold Papers in Russian, East European and Central Asian Studies, Paper No. 12 (The Henry M. Jackson School of International Studies, The University of Washington: May 1997); Daskalova, *Bulgarskiat uchitel prez vuzrazhdaneto* (Sofia: Sofia University Press, 1997); Daskalova, ed. *Ot siankata na istoriata: zhenite v bulgarskoto obshtestvo i kultura* (Sofia: LIK, 1998); Daskalova, *Gramotnost, knizhnina, chitateli i chetene v Bulgaria na prehoda kum modernoto vreme* (Sofia: LIK, 1999); Daskalova, "Women, Nationalism and Nation-State in Bulgaria (1800–1940s)," in *Gender Relations in South Eastern Europe: Historical Perspectives on Womanhood and Manhood in 19th and 20th Century*, ed. Miroslav Jovanovic and Slobodan Naumovic (Belgrade: Udruženje za društvenu istoriju, 2002), 15–37; Daskalova, "Gramotnost," in *Bulgarska kniga. Encyclopaedia*, ed. Ani Gergova (Sofia-Moskva: PENSOFT, 2004), 130–132; Daskalova, "Balkans," in *The Oxford Encyclopedia of Women in World History*, ed. Bonnie G. Smith, vol. 1 (Oxford: Oxford University Press, 2008), 185–195.
15. Reneta Roshkeva and Nadezhda Tsvetkova, *Uchilishteto na Notre-Dame de Sion in Rousse (1897–1948). Dvuezichen album/L'école française Notre Dame de Sion à Roussé (1897–1948). Album bilingue* (Rousse: Avangard Print, 2008).
16. Fournaraki, *"Institutrice, femme et mère."*
17. *Anotero scholeio* is the Greek term for "Hellenic School for Girls," that is, high school for girls. As both Hellenic Schools for Girls and Hellenic Schools for Boys were not state schools, their curricula were somewhat different. According to the 1837 Educational Act, the curriculum of the Hellenic Schools for boys included Ancient Greek, Orthodox Religion, Geography, History, Arithmetic, Calligraphy, [first notions of] Physics and Natural History, French Language. Information provided by Katerina Dalakoura, however, shows that the subjects taught in the Hellenic School for Girls in Hermoupolis in the late 1830s, for example, were the same as those included in equivalent boys' schools.
18. Fournaraki, *"Institutrice, femme et mère."*
19. Ibid.
20. Kavallierakis, *Éducation et écoles étrangères.*
21. Fournaraki, *"Institutrice, femme et mère."*
22. *Parthenagogion* is a general term for both primary and secondary girls' schools. The specific term, *Anotera parthenagogeia* (high schools for girls) names the secondary schools for girls. In reality it was not used for all girls' high schools, so historians must study each school individually to understand what kind of school it really was. *Anotera parthenagogeia* had at least three departments/units—nursery school, primary school, and secondary school. Some

included a separate teachers' training department; department for sewing and needlework, and so on, if finance allowed. Each school/department had its own head, but the general head was the director of the secondary school. With thanks to Katerina Dalakoura for detailed information about differences among various Greek schools during the nineteenth century.

23. Alexandra Bakalaki, "Gender-Related Discourses and Representations of Cultural Specificity in Nineteenth-Century and Twentieth-Century Greece," *Journal of Modern Greek Studies* 12 (1994): 75–112, esp. 79–81.
24. Ziogou-Karasterghiou, "Demoiselles sages et mères parfaites," 399.
25. See Neda Bozinovic and Latinka Perovic.
26. Daskalova, "Women, Nationalism and Nation-State," 23.
27. Daskalova, *Gramotnost, knizhnina*.
28. Daskalova, "Women, Nationalism and Nation-State," 24.
29. *Istoria na obrazovanieto*, 261.
30. Daskalova, *Bulgarskiat uchitel*, 198–215.
31. For more biographical details about Tosheva, Milkova, Miladinova-Alexieva, and Kirkovich, see Nikolai Gentchev and Krassimira Daskalova, eds. *Encyclopaedia, Bulgarska vuzrozhdenska intelligentsia* (Sofia: Petur Beron Press, 1988); for Karavelova, see De Haan, Daskalova, and Loutfi, *A Biographical Dictionary*, 230–234.
32. The Mount Holyoke Female Seminary was founded in 1837 by Mary Lyon in South Hadley, Massachusetts. It was an innovative educational institution for women that trained them as teachers and missionaries.
33. Barbara Reeves-Ellington, "A Vision of Mount Holyoke in the Ottoman Balkans: American Cultural Transfer, Bulgarian Nation-Building and Women's Education Reform, 1858–1870," *Gender & History* 16, 1 (2004): 152–153.
34. Cited in Reeves-Ellington, "A Vision of Mount Holyoke," 158.
35. The statistics come from a large-scale research project on the Bulgarian intelligentsia during the so-called National Revival, comprising roughly the period of the nineteenth century until liberation from Ottoman domination (1878), conducted from 1983 to 1988 at the Centre for Theory and History of Culture, University of Sofia, by a team of researchers, including myself; see Gentchev and Daskalova, *Encyclopaedia*; Daskalova, *Bulgarskiat uchitel*, 223–224.
36. The expression "Bulgarian schools" here refers to the publicly supported schools prior to 1878 and the state schools and publicly supported schools after 1878. No primary schools before Zhivkov's law of 1891 received any financial support from the State. All Bulgarian secondary institutions for boys and girls received state support. Private schools started to open in the Bulgarian towns immediately after the establishment of the nation-state (in some cases even before that, in the 1870s). They existed thanks to the financial resources of foreign, state, private, religious, or secular authorities.
37. *Istoria na obrazovanieto*, 35.
38. Ibid., 11.
39. Ibid., 75.
40. Daskalov, *Bulgarskoto obshtestvo*, 354.
41. *Istoria na obrazovanieto*, 112.
42. Daskalov, *Bulgarskoto obshtestvo*, 370.
43. Ibid., 375–376.
44. Ibid., 378.
45. Ibid., 375–376.
46. Ibid., 377.
47. *Istoria na obrazovanieto*.
48. *Suvmestno obuchenie na momicheta I momcheta*, in *Sbornik s otbrani okruzhni ot Osvobozhdenieto do kraia na 1942 g.*, ed. Nikola Balabanov and Andrei Manev. vol. 1 (Sofia: Ministerstvo na narodnoto prosvieshtenie, 1943), 223–234.

Chapter 11

From an Exclusive Privilege to a Right and an Obligation: Modern Russia

E. Thomas Ewing

In 1897, more than 100 years after the first girls' schools opened in Russia, just 1 percent of the total female population had a completed general secondary education.[1] Seven decades later, 62 percent of the female population aged 16 to 60 had at least an eighth-grade level of education.[2] Over the course of 200 years, from the first girls' secondary schools in the eighteenth century through the educational reforms of the late nineteenth century and following the mass enrollment campaigns of the twentieth century, the number of women earning secondary education increased in absolute numbers, but especially as a proportion of the total female population. A level of educational attainment that began in the eighteenth century as an exclusive privilege of a tiny elite had become, in just 200 years, an achievement not only available to, but expected of and indeed legally obligated from, every girl in this society.[3]

As these trends indicate, the history of girls' secondary schooling in Russia involved a process of expanding access as well as changing expectations within education and accepted practices in society. The history of girls' secondary schooling has been overshadowed in Russian historiography by accounts of elementary schooling, of higher education, and of women's roles as professionals and public intellectuals.[4] Girls' secondary schooling also tends to be eclipsed in policy discussions of curriculum, standards, and sequence, in which boys' schools or male expectations were presumed to represent common needs and expectations.[5] Exploring the history of girls' secondary schools as a distinct subject thus provides further insights into the transition from exclusivity and privilege to rights and opportunities.[6]

Adopting a "gender lens" to view girls' secondary schooling demonstrates the powerful and persistent connections between family roles, employment opportunities, health and social welfare, and the distribution of power in society.[7]

Positioning the history of Russian girls' secondary education in a global framework, beyond the comparisons with Europe that have shaped existing historiography, introduces new measurements that illustrate connections between
secondary schooling and girls' life opportunities and expectations. The topic of
girls' secondary schooling is well suited to such an analytical approach, as current
figures on access, enrollment, and completion illustrate considerable achievements and persistent constraints. According to a 2008 UNICEF report, for
example, while enrollment ratios at the secondary level were approximately equal
for industrialized countries (93 percent of boys and 92 percent of girls enrolled in
secondary schools), the level and the gender ratios showed more variation in the
least developed countries, where secondary schools enrolled 33 percent of boys
and 28 percent of girls. Within the republics of the former Soviet Union, the
enrollment rates represented a narrower range, but primarily at the higher end of
scale. Secondary schools enrolled 97 percent of girls in Kazakhstan, 93 percent
in Uzbekistan, 91 percent in the Russian Federation, 87 percent in Kyrgyzstan,
and the lowest level among the former Soviet republics, 74 percent, in Tajikistan.
While these figures testify to significant regional and gender disparities within
the regions of the former Soviet Union, even the lowest rate was higher than the
average for the Middle East and North Africa, where just 68 percent of girls were
enrolled in secondary schools.[8]

To explain these enrollment patterns, a new approach to gender and education
focuses attention on the factors that keep girls from advancing beyond primary
grades. While regional differences are important, certain common factors do appear
in these studies—as well as "best practices" that can overcome these obstacles. High
attrition rates in primary grades mean that fewer girls are prepared to enter secondary schools. School fees and the indirect costs of educating children rather than
having them contribute to the family income represent economic factors that keep
girls out of secondary schools. Girls are also more likely to be kept out of secondary
schools due to concerns about modesty and appropriate gender roles, which may
be tied to interactions with male teachers and pupils in classrooms, to the need to
travel distances to attend school, or to the obligation that adolescent girls have male
relatives escort them in public. These contemporary studies inform this historical
analysis by identifying specific changes that led girls and their parents to see education as an opportunity rather than a threat to cultural values, a loss of family
income, or a danger to individuals.

In this chapter, secondary schooling refers to grades between elementary level
(usually three or four years) and higher education. These parameters defining
secondary schooling developed gradually, from six years of schooling in the late
nineteenth century to the ten-year school of the Soviet era. Vocational training,
which educated a substantial number of young people at the postprimary level,
is omitted because its close connections to both production and adult education represent substantially different experiences and institutional locations for
individual pupils.[9] By focusing on trajectories in girls' secondary enrollment,
this chapter positions the Russian experience in the broader context of ongoing
efforts to make secondary schooling into a universal right for girls in a global
context.

From Enlightenment to Reaction

In eighteenth-century Russia, education of any kind was an exclusive privilege of a tiny minority of the population. Within this context, however, certain principles prepared the way for future expansions in girls' access to education. In fact, prominent leaders raised the principle of equal educational access between the sexes long before such a goal was attainable in practice or permitted as a policy. While slow in their implementation, these principles would eventually be used to legitimize the expansion of female education in subsequent centuries, even as the assumptions of exclusive privilege yielded to a broader understanding of the advantages of expanding access to secondary schools.

When Empress Catherine II came to power in 1762, Russia had approximately twenty-five secondary schools, educating 6,000 male pupils. Specialized schools created by Peter I in the early 1700s prepared young men for military or government service. In the 1750s, secondary gymnasia began to train boys, from noble as well as non-noble backgrounds, for further study at the Academy of Sciences in St. Petersburg and Moscow University. Only twenty students a year completed these gymnasia, illustrating how small-scale these efforts remained. Elite families also employed tutors for sons, and to a lesser extent for daughters, an educational alternative that only reinforced the exclusivity associated with any kind of advanced schooling.[10] Despite the obstacles posed by cultural values and family roles, a few exceptional women sought some level of advanced education, usually with a degree of encouragement from male relatives.[11]

While Catherine II inherited an educational system structured on principles of preparing an elite for state service, she was also influenced by Enlightenment thinkers to believe in education's potential to transform individuals into better citizens.[12] In a 1767 decree, she proclaimed that the purpose of educating children was "to instill all those Virtues and Qualities, which join to form a good Education; by which, as they grow up, they may prove real Citizens, useful Members of the Community, and Ornaments to their Country."[13] Catherine's approach to education embodied the tension between a social system that privileged masculinity and an idealized vision in which the roles of citizens and subjects were expected of all men and women. In fact, the most influential tract on education from the late eighteenth century, which Catherine coauthored with her advisor, I.I. Betskoi, was entitled: "A general statute for the education of the youth of both sexes."[14] This document called for an education system that would protect children and young adults, age five to twenty-one, from the harmful influences of both families and societies. Rejecting a narrow professional or vocational track, the tract defined the purpose of education as the creation of good citizens and accomplished human beings, with a particular emphasis on moral education. Further proposals for a national system of education described a sequence of schools, open to all social classes and both boys and girls, which taught a curriculum of academic subjects as well as offering moral, religious, and handicrafts training, with justifications grounded in Enlightenment rather than Christian rhetoric. Catherine's views on education reflected an Austrian and Prussian emphasis on educating quiet and useful citizens obedient to the State and accepting of a static and hierarchical society, as opposed to the Lockean or Roussean vision of educating an all-around person.[15] By articulating a vision of schooling for both sexes, Catherine bequeathed a legacy that promoted the principle of coeducation across an emerging school system.

In addition to this principle of educating both sexes, the most tangible legacy of the eighteenth century was the creation of girls' elite boarding schools. First established in the 1760s, Institutes for Noble Girls educated several hundred pupils by the end of the century. These institutes narrowly restricted enrollment to pupils from noble and priestly families.[16] The curriculum focused on preparing girls for familial responsibilities as wives and mothers, with very limited exposure to sciences or any kind of formal education.[17] French conversation and music accounted for a large proportion of the lesson content, and the daily routine emphasized deportment and obedience as the desired skills and attributes of future wives and mothers.[18] Addressing students at the Smolny Institute in 1804, the Russian Empress encouraged young women to fulfill their assigned roles in families and society: "as daughters, to be obedient and respectful; as wives, to be faithful, virtuous, tender, modest, diligent, and useful, and to promote the honor of the person with whom you have linked your fate;...and as mothers, to try to combine warmth toward your children with sensible concern about their future well-being."[19] According to a former pupil, the headmistress of the Smolny Institute, Mme. Lafond, "directed this establishment for thirty years with rare wisdom," to ensure that those in the school fulfilled obligations "conscientiously."[20] While enrollment remained tightly restricted to elite families, these institutes offered the first state-endorsed education for girls beyond the elementary level.[21]

Both the level of enrollment and the quality of instruction in noble schools for girls remained far below the idealized statements of Catherine II. By 1800, less than 2,000 girls studied in the schools. This number represented 8 percent of enrollment in secondary education , as boys outnumbered girls by a ratio of 12 to 1. Most schools offered just three grades of instruction. High attrition rates created obstacles to completing even this basic level of schooling. One-half of all students quit during or after the first year, and just 10 percent completed the three-year course of study.[22] By the end of the reign of Tsar Alexander I in 1825, female enrollment in secondary schools reached 6,000 pupils. Only the elementary grades admitted girls, as secondary and higher education (other than the Institutes for Noble Girls) remained exclusively male. A few girls were admitted, however, either because the administration tolerated coeducation or, in some unusual cases, because of legal ambiguity, as the decrees authorizing schools did not specify the sex of pupils. Yet when the accession of Nicholas I to the throne brought a shift to more conservative educational policies, new legal restrictions on girls' enrollment closed these loopholes that allowed girls to make up a small proportion of enrollment.[23]

The century that followed Catherine's reign thus saw significant changes in girls' schooling. Through proclamations and legislation, the government defined the education of girls as an important civic responsibility and political obligation. According to the statute issued by the "women's patriotic society" in the 1820s, the purpose of "moral education" in their girls' school was "to teach the students to be good wives, solicitous mothers, and exemplary mentors for their children, and to teach them the skills they need to be able to provide for themselves and their families by means of their own hard work."[24] Instructions issued to administrators at girls' schools in the 1850s included these recommendations:

> Since woman is a delicate creature who is naturally dependent on others, her destiny is the family. She should learn that her fate is to submit to her husband, and not to command. She can only ensure her happiness and acquire the love and respect of others... by strictly fulfilling her family duties.[25]

Yet, restrictions on girls' schooling and, more important, the tiny fraction of the female population enrolled in schools ensured that education, and in particular secondary education, remained an exclusive privilege. As the next decades would reveal, substantial progress toward girls' secondary education depended on a combination of government reforms, the mobilization of public opinion, and particularly changes in girls' expectations of schooling's potential benefits.

From Reform to Mobilization, 1850s to 1917

Following the accession of Tsar Alexander II in 1855, the Russian government initiated new educational policies. As a sign of this new approach, Minister of Education Norov recommended expanded education of girls:

> The vast system of education in Russia has up to the present only had in view one half of the population—namely, the males. It would be of greatest benefit to our country to establish girls' schools in provincial and district towns and even in large villages.[26]

This ideal, combined with a strong service ethos across the Russian system of education, found expression in 1858 in Ministry of Education regulations specifying that schools: "impart to pupils that religious, moral, and intellectual education which is expected of every woman, especially the future mother of a family."[27] A draft Statute issued in 1862 proclaimed that "the profits of education ought to be enjoyed by all persons, irrespective of their sex or origin."[28] Despite these ideals, however, female enrollment remained low. In the 1860s, just 0.1 percent of the female population attended any kind of school, compared to 1.3 percent of the male population.[29] Although conservative reaction tempered some reform impulses, this era established the policy of making state elementary schools accessible to boys and girls, in either coeducational or gender-segregated programs, administered by local authorities, and with minimal fees for pupils and their families. While these policies marked an important step toward expanding primary schooling, very few students were prepared for secondary education, due to both the limited curriculum and the small number of pupils completing the primary grades.[30]

The reform era also saw the creation of the first state secondary schools for girls. A government decree in 1858 created two types, with either a three-year or a six-year course of study. In striking contrast to the Institutes for Noble Girls, these state schools enrolled pupils from all social classes.[31] During the 1860s, the curriculum was enlarged to include more subjects, pedagogy was added as a separate field of study, and enrollment broadened to draw from more social groups.[32] Within a decade, 125 girls' schools educated more than 10,000 pupils, testifying to the rapid expansion characteristic of the reform era. Yet these schools received little funding from the State, indicating the low priority the government assigned to girls' education, and thus they depended heavily on local resources.[33] In 1870, most of these schools were renamed as either *gymnasia* or *progymnasia*, following the 1864 statute for boys' secondary schools.[34]

Educational reforms in the 1860s were fueled by both dissatisfaction with Russian social and economic conditions and the sense of hope and possibility nurtured by government reforms, especially the emancipation of the serfs in 1861.[35] Yet the poor quality and limited number of girls' schools also prompted reform-minded educators to press for further changes that would directly affect access to secondary education. The Institutes of Noble Girls provided the object of such criticism, as their narrow curriculum, stagnant pedagogy, and restricted enrollment symbolized the old order in secondary education and gender roles that reformers wanted to eliminate. The most influential criticism came from physician Nikolai Pirogov, who complained in 1856 that these institutes served primarily to turn girls into dolls, while offering no meaningful preparation for adult life, as pupils learned to perform like puppets on a string, rather than act like rational individuals.[36]

During his term as director of the Smolny Institute, noted educator K.I. Ushinskii adopted a stance consistent with Pirogov's criticisms when he told pupils:

> You are obliged to be imbued with the yearning for the conquest of the right to higher education, to make it the goal of your life, to instill that aspiration in the hearts of your sisters, and to attempt to secure the achievement of that goal as long as the doors of the universities, academic and other higher schools are not thrown open before you as hospitably as before men.[37]

Such a statement was truly radical in this context, as it sought to use girls' achievements in secondary education as a kind of wedge to force open the gates of higher education. Perhaps not surprisingly, Ushinskii's policies provoked conservative opposition, and he was forced to resign as director of the Smolny Institute in 1862. Despite criticism from reformers, these institutes remained the exclusive privilege of the elite, with daughters of nobility making up 88 percent of enrollment as late as 1880.[38]

The controversies around elite girls' education reflected the emergence of the "woman question" as a topic of public discourse in the 1860s, as progressive intellectuals connected the status of women with other symbols of Russia's perceived backwardness.[39] This debate was to a great extent influenced by Pirogov's essay, "Problems of Life," which argued that education should be preparation for life, not just preparation for service, and that all children had the same abilities and thus should have the same educational opportunities. According to Pirogov, "the thought of educating herself" for service to the family and the nation should "permeate the moral fibre of women."[40] Although Pirogov's changes provoked controversy and resistance, ultimately compelling his resignation, subsequent generations of educational reformers looked to him as "their spiritual father."[41] For the first time, a leading member of the intellectual community focused on girls' education, thus placing this issue on the agenda and unleashing debates in university corridors and the periodical press. While reformers like Pirogov argued that girls who were better educated would become better mothers, for they would know how to raise their children and also participate in public life, these same critics also accepted the limits of female education. Pirogov did not believe that girls should learn about medicine or about social problems.[42] Even these reformers thus saw female education as a way to strengthen, but not transform, women's roles in families.

While intellectuals embraced the principle of more education for women, the tsarist government pursued policies that inconsistently advanced and occasionally obstructed girls' secondary schooling. New regulations issued in 1868 and 1870 categorized girls' schools as *gymnasia* for the first time, extended the course of study to seven years, with an eighth year of pedagogy, and provided some funding from the Ministry of Education. An expanded curriculum included subjects such as religion, Russian language, math, geography, history, natural science, physics, handwriting, dancing, and needle crafts "for practical application in the home." Graduates of the seven-year program could teach in elementary schools or serve as private tutors, while those completing the pedagogy course in the eighth grade could teach primary grades in a girls' *gymnasium*.[43] As these changes were being adopted, enrollment in more than one hundred girls' schools offering the full six grades of instruction exceeded 9,000 pupils. Yet even this substantial expansion in enrollment encompassed only a small fraction of the school-aged female population.

Government policies regarding funding, teaching personnel, and the administration of girls' secondary schools illustrate the contradictions of a conservative state implementing educational reforms. The government offered very limited resources for girls' secondary schools, which meant that tuition and fees paid by pupils' families and donations from individuals and civic organizations had to provide most of the funding.[44] This policy ensured that teachers' salaries remained lower in girls' schools. Men teachers with higher qualifications generally preferred to work in boys' *gymnasia*.[45] This lack of government funding became a source of concern, and advocates of coeducation argued that whereas boys' *gymnasia* received administrative recognition and financial support from the government, girls' *gymnasia* were forced to "live from day to day," depending on private resources for funding.[46]

Despite the conservative foundations of girls' secondary schools, many pupils knew about the shifting currents of Russian public opinion. While headmistresses tried to maintain girls "like hothouse plants, in a quiet secluded atmosphere far apart from real life, in a world created by their own imagination, and as different from reality as it could be,"[47] the actual experiences of secondary schooling led some girls on the path to radicalization. For Ekaterina Kuskova, who attended a *gymnasium* in the 1870s, inspirational teachers became models of engaged learning and critical thinking. The views expressed by one favorite history teacher even provoked the headmistress, who often observed lessons to monitor their content, to repudiate what was being taught in her school: "Children, children! There's been some sort of misunderstanding here." Declaring that the teacher was misrepresenting the views of poet A.S. Pushkin, the headmistress warned: "Girls should pay attention to the poetic aspect of the verse, and not to the content at all."[48] Other women told similar stories of how "the wave of new ideas" permeated their schools, generating a preoccupation "exclusively with one question—the family discord between the old and the young."[49] In the context of the school, this question could pit female students against teachers, school directors, and state authorities, as in the experience of Kuskova. But female pupils sought out potential sources of support within schools, such as the headmistress described by Praskovia Ivanovskaia, who "expressed no indignation over our unprecedented free-thinking." Yet the most powerful force for radicalization combined exposure to outside ideas, independent reading, and the organization of students' self-study circles, which at times led to organized protests of school policies and practices.[50]

The girls' *gymnasia* also faced contradictions related to the kinds of preparation they could offer to pupils. In direct contrast to their male counterparts, girls' *gymnasia* were not designed to prepare students to enter higher education, particularly as women had been banned from attending Russian universities since the 1860s. Even after the ban was lifted, female pupils who successfully completed all eight grades of a *gymnasium* still had to pass additional examinations, which often involved using their own funds to hire a tutor.[51] The fact that girls' secondary schools did not prepare graduates for higher education ironically became part of the argument against allowing women to enroll in universities.[52]

The differences in curriculum and pedagogy led to charges that girls' *gymnasia* did not offer an education equal to that of boys' *gymnasia*, but rather one adapted to "particularities of the female sex."[53] When the government implemented educational reform, requiring subjects such as history, geography, Russian literature, and jurisprudence, these changes applied only to boys' schools, meaning that girls' schools did not have any minimum educational requirement in these fields. According to one critic, this policy implied that girls in secondary schools should be satisfied with a program deemed antiquated for a boys' school. While some advocates of expanded education for girls praised the *gymnasia* for offering new opportunities for female pupils, which might in turn lead to truly coeducational programs, other critics complained that restrictive administrative policies, curricular limitations, and traditional roles meant that girls' *gymnasia* "lagged behind" the changing conditions of Russia and had even reached "a point of prostration."[54]

Despite the tensions embedded in the structures, policies, and practices of schools, this same period saw substantial increases in enrollment. The final decades of the Tsarist Empire laid the foundation for the massive expansion in education in the twentieth century. By 1914, enrollment in girls' secondary schools exceeded 300,000 pupils, as expanded facilities, rising expectations, and growing demand eroded the exclusive status of privileged institutions. The most significant increases took place in the decades from 1893 to 1913: enrollment doubled from 65,000 to 137,000 by 1903, and then more than doubled again, to 303,000, by 1913. By the early 1900s, the number of girls enrolled in secondary schools exceeded the number of boys, because of the greater opportunities for the latter in vocational and professional education.[55] While this enrollment still represented a small fraction of school-age girls, it nevertheless marked a dramatic expansion, thus illustrating how changes in policies and attitudes could produce significant results despite the obstacles and tensions that persisted in girls' secondary schools.

The distribution of pupils by social class confirms that assumptions of privilege were gradually yielding to perceptions of rights and obligations. Although technically open to all social classes, girls' *gymnasia* charged tuition and fees that, with the indirect costs of educating children, effectively excluded daughters of the poorest classes.[56] In 1904, approximately one-third of pupils in girls' secondary schools were the daughters of nobility or government officials, almost one-half were the daughters of prominent urban families, and the remaining one sixth came from rural social groups, the clergy, and other categories. As enrollment increased, however, so too did the representation of different social groups. By 1914, daughters of nobility and officials made up just 22 percent of enrollment, daughters of leading urban groups had decreased slightly to 44 percent, the daughters of the clergy remained the same at 5 percent, and the main increases in the proportion of students came in the categories of rural social groups and other, which rose to almost 29 percent of the total. Important as well was the significant growth in enrollment of Jewish girls, who, as of 1911, were about 13 percent of the students in state *gymnasia*.[57] While girls' secondary schools illustrated persistent social stratification, as most pupils came from wealthier families and relatively few came from working families, these notable changes in proportions suggest that seemingly exclusive schools were also affected by broader processes of social change.[58]

In the context of Tsarist Russia, the female *gymnasium* student became a powerful symbol of cultural transformation. In the words of S. Kropotkin, a contemporary observer, "The girls' *gymnasia* opened a new era for Russian women." Subjects were taught "in a serious and attractive way" by men with higher educations: "The girl's brain was really working." Recruiting pupils from different classes encouraged "a democratic spirit" in the schools, which found expression even in the clothing of pupils, as "the modest uniform—a brown woolen frock and a little black alpaca apron" became the symbol of a "quite new sort of girl," who longed for more education, "and often for an independent life."[59]

The last decades of Imperial Russia thus saw significant changes in attitudes, policies, and enrollment that indicated how the ideal of education open to both sexes, while still far from fully implemented, had begun to shape trajectories and experiences. The view of education as a privilege of an exclusive elite defined by gender as well as social class, religion, ethnicity, and especially state service yielded to a vision of education as a tool of development, a virtue of citizenship, and a marker of identity. Although further progress would be delayed by a decade of war, revolution, and civil war, the expansion of Soviet girls' secondary schooling built upon the foundations laid before the revolution.

From Revolution to Stabilization, 1917–1970s

In the summer of 1918, less than one year after the October revolution brought the Bolshevik Party into power and established the first communist government, the Soviet State decreed that all educational institutions would be coeducational. Boys' and girls' secondary schools even offered priority enrollment to pupils of the opposite sex, until the number of pupils was approximately equal.[60] This action sought to eliminate the gender hierarchies that progressive educators believed under-lay the conservative foundation of the old regime schools, while also encouraging the egalitarian social relations and acquisition of vocational skills deemed essential to communist society. The principle of equal education for the sexes at all levels of the school system was, for the first time in Russian history, codified in law, even if enrollment patterns and classroom practices continued to reveal profound differ-ences in the education of boys and girls.[61]

The first decade of Soviet power saw only gradual expansions in enrollment of all pupils, but especially girls. The costs of war and civil war, the turmoil of establish-ing a new political system, and the confusion, indirection, and resistance associated with progressive school reforms limited the planned expansion in enrollment in the higher grades.[62] Between 1914 and 1928, the number of pupils enrolled in level II schools (which offered instruction in seven grades) rose from less than 1,000,000 to just over 2,000,000 (by comparison, four-year elementary schools enrolled almost 9,000,000 pupils by 1928). With the introduction of the Five-Year Plans, however, enrollment in schools offering seven or ten years of instruction began to increase dramatically: 3,500,000 in 1930; about 9,100,000 in 1933; almost 15,000,000 in 1936; and more than 20,000,000 by 1939. Whereas enrollment above grade four had been just 400,000 pupils before the war, this same category included 2,000,000 pupils by 1930, more than 6,000,000 by 1935, and more than 10,000,000 pupils by 1939, a fivefold increase in just under a decade.[63] The 1930s thus saw a substan-tial increase in secondary education as the mobilization of educational resources to expand offerings began to achieve significant results with far-reaching implica-tions for the schooling of girls. Responding to this dramatic expansion of enroll-ment, while also seeking to mobilize the population to see secondary education as an obligation as well as a right, the Communist Party proclaimed in 1934 the goal of achieving universal and compulsory education at the seventh-grade level.[64] The educated girl became a powerful symbol of how communism had transformed both gender roles and access to specialized knowledge.[65]

Girls' enrollment in elementary and secondary schools rose from less than 5,000,000 in 1928 to nearly 12,000,000 by 1936. The proportion of girls among pupils also rose significantly in this same period, from less than 40 percent to nearly 47 percent.[66] The proportion of women among secondary teachers reached nearly one-half by 1939, meaning that more women served as role models for girls pursu-ing further education, even as this sector of professional employment offered more opportunities for these same girls.[67] For girls enrolled in secondary level classes, like Moscow student Nina Kosterina, women teachers who had been educated in Soviet schools offered a path that combined civic contributions and personal achievement.[68]

But these trajectories also obscured persistent gaps and notable tensions in girls' education, especially at the secondary level. Only one-half of the female school-age population was enrolled, indicating that the goal of universal and compulsory edu-cation was far from realized. Regional differences remained significant, as enroll-ment beyond the primary grades remained lower in rural areas and particularly in the so-called "backward" regions of the Caucasus, the Far North, and Central

Asia.[69] More girls dropped out with each passing grade, particularly in the transition from elementary to secondary grades.[70] While Soviet schools in the 1920s and 1930s were fully coeducational in terms of curriculum and pedagogy, the 1943 decision to introduce separate schools in the major cities revealed that even the principles of gender equality had not eliminated persistent differences in the education of boys and girls.[71]

Despite these continuing tensions, however, girls' secondary schooling expanded at an unprecedented rate during the remainder of the Stalinist period. By 1939, approximately one-third of women in their twenties, those who were of school age during the period of transformation, had completed some level of secondary schooling. In 1955, girls made up almost one-half of all pupils in Soviet schools (49.6 percent), but were actually represented at higher percentages in grades 5–7 (48.7 percent) and grades 8–10 (55.4 percent) than in the elementary grades (47.9 percent). Continuing trends begun in late imperial Russia, but now at an accelerated pace, the rate of enrollment expanded most significantly among the Muslim communities: between 1939 and 1959, the percent of adolescent girls and adult women with at least a secondary education rose from less than 5 percent to approximately 30 percent in the Central Asian republics of Kazakhstan, Uzbekistan, and Kirgizstan. By the end of the 1950s, almost 30,000,000 women had acquired some level of secondary schooling. The proportion of women with at least a seventh-grade education rose from 9 percent in 1939, to 34 percent in 1959, to 45 percent in 1970, and finally reached a majority, 53 percent, by the mid 1970s. As this educational expansion continued, the youngest generation of women experienced the most dramatic changes in their educational accomplishments. When American educational researchers took advantage of the "thaw" that followed the death of Stalin to observe first-hand the practices and policies of communist education, they commented on "the complete equality" between men and women that originated in the schools, where boys and girls treated each other with "dignity and mutual respect."[72] Among the women who began their schooling in the postwar decades of the 1940s and 1950s, more than 90 percent had completed an education beyond the primary school, according to the 1970 census. In 1975, female pupils made up 56 percent of the 6,000,000 pupils enrolled in grades eight, nine, and ten, while making up exactly one-half of the more than 43,000,000 pupils in all Soviet schools.[73]

This rate remained far below the repeatedly proclaimed goal of universal secondary education at the tenth grade level.[74] Nevertheless, the Soviet Union had made considerable progress toward the goal of educational equality for the sexes at the secondary level. The secondary education of girls thus came to be seen as both a right and an obligation. Even those individuals alienated from or victimized by the Soviet government could recognize and even demand the fulfillment of promises of women's equality and educational opportunity. This point was vividly illustrated in the memoirs of physician Vera Ivanovna Malakhovka, who recalled from her own experience that Soviet power "wasn't good in a lot of ways; all our friends were thrown in prison, a lot of people we knew, and in fact, many people at that time suffered terrible fates." Yet when faced with the question of her daughter pursuing further education, Malakhovka's mother never hesitated: "Mama always said that thanks to Soviet power, you could study anywhere you wanted, that anyone who wanted to could study for free—as long as you weren't lazy."[75] The consolidation of girls' rights and obligations at the secondary level occurred during the mobilization campaigns of the Stalinist era, even as the regime pursued repressive strategies that would ultimately contribute to the ideological collapse of communism.

Understanding Secondary Schooling for Girls

Soviet historians emphasize that it was the Communist revolution that brought the most significant changes in women's educational levels.[76] While the revolution of 1917 certainly led to important advances in girls' secondary schooling, so too did the Enlightenment principles endorsed by Catherine II, the establishment of state schools in the early nineteenth century, the reforms and intellectual debates of the 1860s, the mobilization of social forces in support of educational opportunities and women's rights in the 1910s, and the compulsory mobilizations of the Stalin era. As schools enrolled more girls from different backgrounds, secondary education became an opportunity rather than an exclusive privilege; as women entered new fields of employment and as the State made new demands on citizens, secondary education became both a right and an obligation for growing proportions of the female population.

The history of girls' secondary education in modern Russia meant an expansion in rights and opportunities as well as changes in expectations and obligations. As the system of primary education expanded, more girls continued into secondary schooling. As policy-makers, educators, and the public came to accept and then promote secondary education for girls, institutional reform, political intervention, and public opinion facilitated the expansion and transformation of this sector of Russian/Soviet education. Yet at each stage it was girls themselves, including the individual examples cited in this chapter, who actually crossed "the bridge" from primary to secondary schools; these girls were the agents who determined the trajectory and significance of this historical experience.[77] This study thus confirms the significance of the incremental and situational steps recommended by recent studies of girls' schooling.[78] The history of girls' secondary schooling in Russia demonstrates how substantial transformations occur through a sequence of small-scale changes in policies, practice, and perceptions, all of which acquire meaning at the everyday level of the girls involved in these schools.

Notes

1. William Johnson, *Russia's Educational Heritage* (Pittsburgh: Carnegie Press, 1950), 285.
2. *Itogi vsesoiuznoi perepisi naseleniia 1970 goda*, vol. 3 (Moscow: Statistika, 1972), 6–7.
3. For women's rights and educational expansion as markers of modernity, see David L. Hoffmann, *Stalinist Values: The Cultural Norms of Soviet Modernity, 1917–1941* (Ithaca: Cornell University Press, 2003), 62–71, 88–117.

4. For elementary schooling, see Ben Eklof, *Russian Peasant Schools: Officialdom, Village Culture, and Popular Pedagogy, 1861–1914* (Berkeley: University of California Press, 1986); John Dunstan, "Girls and Schooling in Imperial Russia. An Outline Mainly from Western Research," *Education in Russia, the Independent States, and Eastern Europe* 14, 1 (1996): 15–17. For women and higher education, see Christine Johanson, *Women's Struggle for Higher Education in Russia, 1855–1900* (Montreal: McGill-Queen's University Press, 1987). For women professionals in education, see Christine Ruane, *Gender, Class, and the Professionalization of Russian City Teachers, 1860–1914* (Pittsburgh: University of Pittsburgh Press, 1994). For radical women, see Barbara Alpern Engel, *Mothers & Daughters: Women of the Intelligentsia in Nineteenth-Century Russia* (Cambridge: Cambridge University Press, 1983).

5. For criticism of policies that ignored girls' schools, see N.A. Skvortsov, "Ocherednye voprosy zhenskago obrazovaniia," *Zhenskaia zhizn'* 16, August 22, 1915, 1.

6. For surveys that include discussion of girls' secondary schooling, see A.G. Rashin, "Gramotnost' i narodnoe obrazovanie v XIX i XX vv.," *Istoricheskie zapiski* 37 (1951): 75–76; Nicholas Hans, *History of Russian Educational Policy (1701–1917)* (New York: Russell & Russell, 1964), 235; Dunstan, "Girls and Schooling," 8–22; L.D. Filippova, "Iz istorii zhenskogo obrazovaniia v Rossii," *Voprosy istorii* 2 (February 1963): 209–218. Enrollment patterns showed considerable variation in the Russian empire, with the highest rates in the western borderlands and especially among Jewish populations, while significantly lower rates of education characterized the Caucasus and Central Asia. See discussion in Wayne Dowler, *Classroom and Empire. The Politics of Schooling Russia's Eastern Nationalities, 1860–1917* (Montreal: McGill-Queen's University Press, 2001).

7. For the "gender lens" in studies of education, see UNESCO, *Education for All. Is the World on Track?* (Paris: UNESCO, 2002), 78.

8. UNICEF, *The State of the World's Children 2008* (New York: United Nations Children's Fund, 2007), 132–133.

9. For professional, technical, and vocational education for girls and women, see John Dunstan, "Coeducation and Revolution: Responses to Mixed Schooling in Early Twentieth-Century Russia," *History of Education* 26, 4 (1997): 378; and "Girls and Schooling," 18–19.

10. Isabel de Madariaga, *Russia in the Age of Catherine the Great* (New Haven: Yale University Press, 1981), 488–489.

11. See accounts in A. Woronzoff-Dashkoff, *Dashkova: Life of Influence and Exile* (Philadelphia: American Philosophical Society, 2007), 12–13; Sonya Kovalevsky, *Her Recollections of Childhood*, trans. Isabel F. Hapgood (New York: Century, 1895), 65–67; Ekaterina Kuskova, "Memoir," in *Russian Women 1696–1917: Experience and Expression: An Anthology of Sources*, ed. Robin Bisha, Jehanne M Gheith, Christine Holden, and William G. Wagner (Bloomington: Indiana University Press, 2002), 184–185.

12. Madariaga, *Russia*, 489–491.

13. Quoted in W.F. Reddaway, ed., *Documents of Catherine the Great: The Correspondence with Voltaire and the Instruction of 1767* (New York: Russell and Russell, 1931), 272–273.

14. I.I. Betskoi, "General'noe ucherezhdenie o vospitanii oboego pola iunoshestva," in *Khrestomatiia po istorii shkoly i pedagogiki Rossii*, ed. S.F. Egorov and Sholom Izrailevich Ganelin (Moscow: Prosveshchenie, 1974), 66–69; Hans, *History*, 18; Madariaga, *Russia*, 491–492.

15. See discussion in Carol S. Nash, "Educating New Mothers: Women and the Enlightenment in Russia," *History of Education Quarterly* 21, 3 (Autumn 1981): 301–316; Madariaga, *Russia*, 496–501; Dunstan, "Girls and Schooling," 10–12.

16. Filippova, "Iz istorii," 214; Hans, *History*, 18–19; Madariaga, *Russia*, 493.

17. Filippova, "Iz istorii," 214.

18. "Anna Virt, Petition to Open a Girls' School," in *Russian Women*, 176–179; Johanson, *Women's Struggle*, 3.

19. Engel, *Mothers*, 24.

20. "Memoirs of Glafira Rzhevskaia," in *Russian Women*, 113.

21. Hans, *History*, 19; Filippova, "Iz istorii," 214; Johanson, *Women's Struggle*, 3.

22. Filippova, "Iz istorii," 212.

23. Hans, *History*, 58.

24. "Women's Patriotic Society, Statute of Girls' School," in *Russian Women*, 181.

25. Engel, *Mothers & Daughters*, 24.
26. Cited in Hans, *History*, 96.
27. Quoted in Johanson, *Women's Struggle*, 32.
28. Quoted in Hans, *History*, 101.
29. Filippova, "Iz istorii," 209; Johanson, *Women's Struggle*, 3.
30. Hans, *History*, 105–107.
31. Johnson, *Russia's Educational Heritage*, 146.
32. Hans, *History*, 97.
33. Johanson, *Women's Struggle*, 29.
34. Johnson, *Russia's Educational Heritage*, 146.
35. Eklof, *Russian Peasant Schools*, 50–69.
36. Johanson, *Women's Struggle*, 4, 10–11. See also criticisms reported in L.E. Raskin, "Bor'ba za devochku kak odna iz problem vospitaniia," *Na putiakh k novoi shkole* 2 (1929): 49.
37. Quoted in Johanson, *Women's Struggle*, 11.
38. Johanson, *Women's Struggle*, 11–12, 20–23, 32, 51–52.
39. Filippova, "Iz istorii," 210–211.
40. Quoted in Engel, *Mothers & Daughters*, 52.
41. Hans, *History*, 98–99.
42. Johanson, *Women's Struggle*, 10–11.
43. "Government Statute for State Women's Gymnasiums and Progymnasiums," in *Russian Women*, 181–183; Hans, *History*, 128; Johanson, *Women's Struggle*, 127–129.
44. Patrick Alston, *Education and the State in Tsarist Russia* (Stanford: Stanford University Press, 1969), 203; Filippova, "Iz istorii," 214.
45. Skvortsov, "Ocherednye voprosy," 1; Alston, *Education*, 203–204.
46. E. Merkhelevich, "Ideiia sovmestnogo obucheniia v istorii nashei shkoly i eia politichecksoe znachenie," *Soiuz Zhenshchin* 3 (1909): 6–7.
47. S. Kropotkin, "The Higher Education of Women in Russia," *The Nineteenth Century*, 43, 251 (1898): 118.
48. Kuskova, "Memoirs," in *Russian Women*, 190–191.
49. Praskovia Ivanovskaia, in *Five Sisters. Women Against the Tsar*, ed. Barbara Alpern Engel and Clifford N. Rosenthal (Boston: Allen and Unwin, 1987), 98–99; Kovalevsky, *Her Recollections*, 93.
50. Ivanovskia and Kovalskaia Memoirs, in *Five Sisters*, 99–101, 209–210; Kuskova, "Memoirs," in *Russian Women*, 185–191.
51. N.A. Skvortsov, "Ocherednye voprosy," August 22, 1915, 2.
52. Merkhelevich, "Ideiia sovmestnogo obucheniia," 7–8.
53. Merkhelevich, "Ideiia sovmestnogo obucheniia," 6.
54. Skvortsov, "Ocherednye voprosy," August 22, 1915, 1.
55. Rashin, "Gramotnost," 75–76; Hans, *History*, 235; Dunstan, "Coeducation," 378.
56. Johanson, *Women's Struggle*, 29.
57. Rashin, "Gramotnost," 76; Sophia Satina, *Education of Women in Pre-Revolutionary Russia*, trans. Alexandra Poustchine (New York: n.p., 1966), 50.
58. Filippova, "Iz istorii," 214; A. Gorchakov and I. Komarov, "Sovmestnoe obuchenie," *Shkola i zhizn'* 10 (1927): 45.
59. Kropotkin, "Higher Education," 120–121.
60. "O vvedenii obiazatel'nogo sovmestnogo obucheniia," May 18, 1918, in *Sobranie uzakonenii i rasporiazhenii rabochego i krest'ianskogo pravitel'stva* 39 (June 4, 1918), 475.
61. Gorchakov and Komarov, "Sovmestnoe obuchenie," 43–47; Raskin, "Bor'ba za devochku," 53–54; and "Eshche o sovmestnom vospitanii," *Na putiakh k novoi shkole* 6 (1929): 55–58. For secondary schools in a revolutionary society, see Larry E. Holmes, *The Kremlin and the Schoolhouse. Reforming Education in Soviet Russia, 1917–1931* (Bloomington: Indiana University Press, 1991), 84–92. For resistance to coeducation, see E. Thomas Ewing, "The 'Problem of Boys and Girls': Gender Relations in Soviet Schools of the 1920s" (unpublished paper, 2008).
62. E.N. Evergetova, *Mal'chiki i devochki v Sovetskoi shkole. K voprosu o sovmestnom vospitanii i obuchenii* (Moscow: Gosudarstvennoe izdatel'stvo, 1928), 8.

63. *Kul'turnoe stroitel'stvo SSSR* (1940), 7, 37; *Molodezh' SSSR. Statisticheskii sbornik* (Moscow: Gosplan, 1936), 217.
64. "O podgotovke k vvedeniiu semiletnogo vseobshchego obiazatel'nogo politekhnicheskogo obucheniia," March 13, 1934, *Narodnoe obrazovanie v SSSR* (Moscow: Pedagogika, 1974), 115–116.
65. See, for example, the striking description of a young educated woman named Nadya as a symbol of Soviet modernization in Maurice Hindus, *Red Bread. Collectivization in a Russian Village* (Bloomington: Indiana University Press, first published 1933, reprinted 1988), 1–9.
66. *Zhenshchina v SSSR* (Moscow: Gosplan, 1936), 108.
67. This estimate is based on figures from *Kul'turnoe stroitel'stvo* (1940), 86, 90.
68. Nina Kosterina, *The Diary of Nina Kosterina* (New York: Avon, 1968), 51–52, 57–58, 84, 93. For women teachers in the Stalinist era, see E. Thomas Ewing, "A Stalinist Celebrity Teacher: Gender and Professional Identities in the Soviet Union in the 1930s," *Journal of Women's History* 16 (Winter 2004): 92–118; "Personal Acts with Public Meanings: Suicide by Soviet Women Teachers in the Stalin Era," *Gender & History* 14 (April 2002): 117–137; "Silences and Strategies: Soviet Women Teachers and Stalinist Culture in the 1930s," *East/West Education* 18 (Spring 1997): 24–54.
69. For campaigns to enroll "non-Russian" girls, see K.I.L'vov, "Po shkolam Kirgizii," *Kommunisticheskoe prosveshchenie* No. 2 (1934): 136; L. Shnaider, "Ni odnoi devochki za bortom shkoly," *Za kommunisticheskoe prosveshchenie*, March 8, 1932, 3; Ewing, "Not One Girl Outside the School." (Unpublished paper, 2007).
70. Raskin, "Bor'ba za devochku," 57.
71. For the Soviet separate schools, see E. Thomas Ewing, "The Repudiation of Single-Sex Schooling: Boys' Schools in the Soviet Union, 1943–1954," *American Educational Research Journal* 43, 4 (2006): 621–650; and *Separate Schools. Gender, Policy, and Practice in Postwar Soviet Education* (manuscript in progress, 2009). For girls' schools, see B. Aleksandrov, "U direktora zhenskoi shkoly," *Uchitel'skaia gazeta*, September 8, 1943, 2; Ia. Pilipovskii, "V zhenskoi shkole. 36-ia sredniaia shkola gor. Sverdlovska," *Uchitel'skaia gazeta*, February 4, 1944, 3; S. Razvinova, "Vospitanie uchashchikhsia zhenskoi shkoly," *Uchitel'skaia gazeta*, August 10, 1949, 3.
72. *Soviet Commitment to Education. Report of the First Official U.S. Education Mission to the USSR* (Washington: U.S. Department of Health, Education, and Welfare, 1959), 2.
73. *Kul'turnoe stroitel'stvo SSSR* (1940), 7; *Kul'turnoe stroitel'stvo SSSR. Statisticheskii sbornik* (Moscow: Gosizdatel'stvo, 1956), 176; *Zhenshchiny v SSSR Statisticheskii sbornik* (Moscow: Statistika, 1975), 61–62, 66; *Itogi vsesoiuznoi perepisi naseleniia 1970 goda*, 3: 6–7; Alastair Macauley, *Women's Work and Wages in the Soviet Union* (London: George Allen & Unwin, 1981), 140–141.
74. George Z.F. Bereday, et al., eds., *The Changing Soviet School* (Boston: Houghton Mifflin Company, 1960), 186–187.
75. Vera Ivanovna Malakhova, "Four Years as a Frontline Physician," in *A Revolution of Their Own. Voices of Women in Soviet History*, ed. Barbara Alpern Engel and Anastasia Posadskaya-Vanderbeck (Boulder: Westview, 1998), 184.
76. Filippova, "Iz istorii," 209.
77. For bridging elementary and secondary schooling, see May A. Rihani, *Keeping the Promise: Five Benefits of Girls' Secondary Education* (Washington, D.C.: Academy for Educational Development Center for Gender Equity, 2006), 11–30.
78. These steps include eliminating or at least substantially reducing school fees, providing safe schools in close proximity to families, addressing the household burdens of girls that often lead to absences and attrition, making schools "girl friendly" by protecting girls while also encouraging them to transcend cultural stereotypes, recruiting women to teaching to provide role models of educated and respected adult women, and offering a quality education with trained teachers, updated textbooks, and a curriculum meaningful to girls and their families. Barbara Herz and Gene B. Sperling, *What Works in Girls' Education: Evidence and Policies from the Developing World* (Washington, D.C.: Council on Foreign Relations, 2004), 45; UNESCO, *"Scaling Up" Good Practices in Girls' Education* (Paris: UNESCO, 2005), 29; UNESCO, *Education for All*, 77.

Chapter 12

Europeans and the American Model of Girls' Secondary Education

James C. Albisetti

By far the most important era for European interest in American education for girls was the period from the end of the U.S. Civil War through the first decade of the twentieth century. Before 1865, there were both fewer visitors to the United States and fewer institutions capable of provoking interest than in later decades; in addition, establishing or significantly upgrading girls' schooling had not yet become a significant issue in much of Europe. Recent studies of ideas about female education in England, France, and Germany from 1760 to 1810 and of the realities of girls' schooling in England and France from 1800 to 1867 suggest that American practices drew no attention at all during these years. Rebecca Rogers has pointed out, in fact, how Catholic teaching orders carried French models of female education to the United States at that time.[1]

By the end of the First World War, almost all European countries had established secondary schooling for girls wanting to pursue higher education and also for those looking to stop short of the secondary diploma for boys. If transatlantic exchanges continued and in some cases expanded, American secondary education in general did not provide much inspiration: "high schools were more criticized than any other part of the educational system" by German pedagogues during the Weimar Republic. Even in occupied Germany after 1945, American models of longer common schooling, a more comprehensive secondary school, and coeducation encountered strong resistance.[2] When many countries in Western Europe greatly expanded secondary enrollments and adopted more widespread coeducation in the 1960s and 1970s, there was less discussion of American models than before 1914.

The Era before the Civil War

In the early 1800s, European interest in American education focused primarily on the common schools, especially as they developed in Massachusetts, Connecticut, and New York. These tax-supported nondenominational schools proved to be of particular interest to English reformers trying to overcome the serious deficiencies of the "voluntary" system of church-supported schools. The phrenologist George Combe, who toured the United States from 1838 to 1840, was the prime source of information about American practices, especially in Massachusetts. Combe's writings served as a basis for much of the agitation by the Lancashire—later National—Public School Association in the late 1840s and early 1850s.[3] Interest in American common schools arose in Italy as early as 1832 among the Florentine reformers associated with the periodical *Antologia*. The most analytical study of American schooling by a European in this era, however, came in 1852 from the Swede Per Adam Siljeström, who devoted 300 pages to the common schools, only ten to secondary education.[4]

Most visitors before the Civil War commented on the numerous young female teachers in the common schools, some of whom taught boys nearly their own age. For Combe, the Prussian Friedrich von Raumer, and Richard Cobden, this situation was not particularly problematic. Siljeström found it, however, "strange, as regards our habits and ideas." Right after the Civil War the German-American Rudolf Dulon strongly criticized having women teach older boys, and the Belgian Protestant Emile de Laveleye claimed no European father would send his son to such a school.[5]

Commentary on academies and seminaries for girls in this era was rare and often ambivalent. Englishwoman Frances Trollope disparaged blushing sixteen-year-old girls being examined in mathematics and moral philosophy in Cincinnati around 1830, yet she admired the "enlarged scale of instruction proposed for young females" at a fashionable boarding school in New York. Combe witnessed public examinations at the Albany Female Academy in 1840 and found the girls' "attainments were highly creditable to themselves and to their teachers." He also heard, though, that "they are stimulated to excess by emulation, and that they occasionally ruin their health by exertions to gain prizes." A decade later, Frederika Bremer of Sweden found the Rutgers' Female Institute in New York to have "excellent arrangements for their instruction and cultivation"; but she worried that publishing girls' literary works would lead to harmful "ambition." After inspecting the Female Academy in Brooklyn, Bremer wondered whether "these finishing schools for young girls...develop that which is best in woman? I doubt it."[6] None of the three suggested that Europeans should imitate these American models.

The Post-Civil War Era

After 1865, Europeans' general interest in the United States increased dramatically. In part, improvements in travel across the Atlantic made visiting easier for both tourists and investigators sent by governments. The international expositions held in Philadelphia in 1876, Chicago in 1893, and St. Louis in 1904 drew significant waves of visitors. The majority saw primarily the northeast quadrant of the country, bounded by Boston, Washington, St. Louis, and Chicago, with the obligatory detour to Niagara Falls.[7] During this era the image of the United States changed from that of a "backward" frontier country to one of an urbanizing and industrializing giant, which one German visitor in 1903 called the "land of the future."[8] Contributing significantly to this metamorphosis was the sense that if the position of women was an important measure of the advance of civilization, then the United States had moved ahead of Europe.

Two important English studies of American education appeared shortly after the Civil War. Rev. James Fraser, later the Bishop of Manchester, investigated public elementary and secondary schools in the United States for the Schools Inquiry Commission, with the special charge to study the role of women teachers and "the degree in which both sexes and different ranks in life are associated in the same school." Sophia Jex-Blake toured privately, visiting pioneering coeducational colleges such as Oberlin, Antioch, and Hillsdale as well as high schools and normal schools.[9]

In France, Édouard Laboulaye's *Paris in America*, published in 1863, highlighted female physicians and their medical schools, secondary coeducation, and women teachers. Longtime Parisian resident and hostess Mary Clarke Mohl wrote in December 1867 that Laboulaye "told us that during the last few years, experiments have been made in America, at certain colleges, to bring up girls and boys in exactly the same way, and their faculties were exactly equal, varying according to the individual, apart from the sex."[10] In an article in the *Revue des deux mondes* in 1869, Célestin Hippeau linked advances in the education of women and of freed slaves. He insisted that the United States led most of Europe in female education, noting that coeducation was "a question that could not even be posed for France."[11]

Interested Italians learned of American developments largely through French sources, including a translation of Laboulaye's book. In 1869 the Neapolitan educator Edoardo Fusco emphasized the wide-ranging studies undertaken by American girls and women, and suggested that "the part they play in the social and especially pedagogical progress of their country is something not easily believed in Italy." The following year, the influential pedagogue Aristide Gabelli drew heavily on Hippeau's work for an article on female education in *Nuova Antologia*.[12]

Dulon's book brought both single-sex and coeducational academies to German attention in 1866. By 1872, Gotthold Kreyenburg, director of a municipal girls' school in Iserlohn, displayed broad familiarity with—and a general distaste for—conditions in the United States in a comparative study of girls' education. In Austria, awareness of American developments owed much to the materials sent by the U.S. Bureau of Education to the Vienna Exhibition in 1873, including a short history of Vassar College written by its president. The Centennial Exhibition in Philadelphia spurred the first serious study of American schooling by an Austrian, Franz Migerka.[13] The numerous other official and unofficial accounts produced after the Centennial Exhibition meant that information about girls' schooling in the United States was available to most Europeans who cared to seek it.[14]

What Qualified as American Secondary Education?

Historian William Reese has argued that "educators North and South as late as the 1870s realized that there was really no such thing as *the* American high school." Given that almost all European secondary schools for boys began before age twelve, that many followed private elementary classes, that graduates were often nineteen or twenty, and that even the Girls' Public Day School Company in England allowed attendance to age twenty, the four-year American high school that followed an eight-year common school appeared to many visitors as less than a true secondary school.[15]

Many Europeans stressed the close ties of high schools to elementary education. Several Frenchmen, including the important official Ferdinand Buisson, compared them to the new higher primary schools in their own country. In a similar fashion, Joshua Fitch, an inspector of English training colleges, claimed in 1890 that a high school was "essentially a continuation school." The broad social mix of students, coeducation, and the prevalence of women teachers all contributed to such negative perceptions. Especially for male visitors from England and Germany, the third factor became the key problem, with regard to both academics and the upbringing of "virile" young men. German-American L.R. Klemm shared the distaste for women teachers, but in an unusual stance suggested that the average male graduate was much better prepared "to find his way in life" than his older German counterpart completing the *Gymnasium*.[16]

Europeans who considered high schools to be "secondary" nonetheless highlighted their limited range. Dulon claimed that a German *Gymnasium* achieved more in three years than an American academy did in four or five. Some observers compared the high school to a Swiss or German *Realschule* that ended at age sixteen, or to the lower grades of a French *lycée*. As late as 1909, Sara Burstall, headmistress of the Manchester High School for Girls, asserted, "The high school course begins too late; twelve, not fourteen, is the proper age for secondary education to begin."[17]

Most European observers included American colleges as elements of "secondary" education. French pedagogue Gabriel Compayré even claimed in 1896 that Harvard College was "a school of secondary education, where one prepares for the bachelor of arts." Siljeström had suggested forty years earlier that the American college resembled a *Gymnasium*, not a university. In the 1860s Jex-Blake compared the social origins and studies of students at Oberlin to those in English training colleges. In European eyes, the proliferation of American colleges also made them seem like secondary schools. Ohio alone had thirty-two degree-granting colleges; for English historian Edward A. Freeman, the diplomas "of some of these institutions cannot be worth very much."[18]

European discussions of secondary schooling for American girls thus included both coeducational colleges and universities and the new women's colleges that emerged after the Civil War.[19] Among the latter, Vassar (opened 1865), Wellesley (1875), and Bryn Mawr (1884) drew the most attention, located as they were in easy reach of major eastern cities. Smith (1875) and Mt. Holyoke (raised to a college in 1887) attracted fewer visitors. News of Vassar appeared in England before it opened, with two articles in the *English Woman's Journal* labeling it first a "Female College" and then the "American University for Women." Hippeau, the first important foreign visitor to Vassar, claimed that there "the right of women to superior instruction has been recognized." Writing in 1871, German pastor Philipp von Nathusius, an opponent of women's emancipation, nonetheless accepted that Vassar was a "Ladies' University."[20]

More skeptical voices soon arose. Kreyenburg argued that the admission age of fifteen at Vassar meant its students were barely older than German elementary school graduates. James Orton, professor of natural history at Vassar from 1869, told an English audience in 1874 that it was "neither a seminary nor a university." Even Bryn Mawr, the only women's college that offered graduate training from its foundation, appeared to a Swiss visitor in 1894 as "secondary education." Frenchman Charles Barneaud compared it to the normal schools for women secondary and elementary teachers at Sèvres and Fontenay, which certainly ranked well below universities. For Burstall, the women's colleges in the 1890s seemed "rather to resemble boarding schools. The whole morning is taken up with recitations, as in a school."[21]

German novelist Ludwig Fulda viewed those attending women's colleges as "pupils" (*Schülerinnen*), not students. In a similar fashion, Italian Carlo Gardini spoke of "boarding school girls" (*educande*) at Vassar. Migerka also referred to the students as "girls," but he was one of the first foreigners to recognize that the top women's colleges "in part extended beyond" secondary education.[22] Awareness of the differing quality of American institutions grew by the end of the century. The same year that Bryn Mawr was labeled as "secondary education," German Emil Hausknecht counted it among the eleven "genuine" universities in the United States. In the early 1890s as well, one article published in France indicated that Wellesley combined secondary and higher education, while another labeled the top women's colleges clearly as "higher" education.[23]

Appealing to American Models

Reformers in many countries referred to American coeducation and/or women's colleges to support demands for improved educational opportunities, though few wanted to imitate the American high school. As early as 1868, women in St. Petersburg took inspiration from Vassar in framing petitions for a women's university. Two years later in the Netherlands, Marie Delsey cited a strong defense of coeducation by Oberlin's president, James Fairchild, in her argument for admission of Dutch women to secondary and higher education. A similar use of the success of coeducation in the United States came in 1871 from a surprising source, Maria Grey, who a year later helped to found the Girls' Public Day School Company. Grey told the Society of Arts that mixed schooling at all levels "would be an advantage to both sexes, morally as well as intellectually." Only ten years later did she make explicit that much of the "weighty evidence" she relied on came from the United States, where coeducation "has been fully tried not only in schools but in colleges." In Catholic Austria, during discussion of the opening of the first *Mädchenlyzeum* in

Graz in 1873, a member of the provincial school council lamented how his country did so much less for female education than "America or even Russia."[24]

In the mid-1870s in Spain, Don Francisco Giner de los Rios referred frequently to American institutions in the prospectus and first reports of the coeducational *Institución Libre de Enseñanza* in Madrid. At the same time, the munificent philanthropy of Matthew Vassar provided a direct impetus to wealthy English businessman Thomas Holloway, although his creation, Royal Holloway College, did not open for more than a decade.[25]

Delegates to the first international feminist congress in Paris in 1878 passed a resolution in favor of coeducation at all levels of schooling that included the claim that American experience had shown it to be "a powerful stimulant for the progress of studies and of moralization." In the report that Camille Sée presented two years later to the Chamber of Deputies in support of his bill creating *lycées* for girls, he placed evidence from the United States first. In typical fashion, he included information on Vassar and Wellesley in addition to the single-sex and coeducational high schools of Boston. Yet the single-sex schools stopping at age sixteen that he proposed bore little resemblance to American models. When questioned in the legislature why he did not propose coeducation, given the great attention it received in the report, Sée responded, "I know very well that France is a Catholic country that does not have the mores of Protestant countries, where they apply coeducation of the sexes."[26]

Appeals to American models were less frequent in Germany, a country that sent few women travelers to the United States. Yet Helene Lange, the leading figure in the struggle for improved secondary schools and admission of women to higher education in the late nineteenth century, recalled in her memoirs that she learned during the 1880s that in the United States, "boys' and girls' education was exactly the same and the number of women teachers greatly exceeded that of men."[27]

In England in the mid-1890s, the Bryce Commission solicited information about coeducation from several leading American educators and commissioned reports from two British visitors, Alice Zimmern and J.J. Findlay. Across the Irish Sea in Dublin the Association of Women Graduates bolstered their call for coeducational higher education in 1902 with "a letter from W.T. Harris, the U.S. Commissioner of Education." In a similar fashion, when someone suggested at the All Russian Women's Congress in 1908 that coeducation would lead to flirting and philandering, she was immediately "refuted from the experience of the U.S.A. and Finland."[28]

Rejecting the American Model

Belief in traditional gender roles, often reinforced by religion, enabled some Europeans to dismiss American practices despite their successes. Fraser ended his discussion of secondary coeducation with a reference to "He who 'at the beginning made them male and female.'" Catholics almost universally criticized coeducation and many rejected similar secondary curricula for girls and boys. For example, the Belgian Redemptorist Francis Xavier Godts condemned the "American innovation of coeducation" and insisted, "In Europe ... we will never raise our seminarians, our future soldiers and sailors, like girls, nor our girls like them."[29]

One way to disparage the success of American coeducation was to point to the "easy" curricula in high schools and colleges. Many Europeans thought that girls would never keep up with boys if they faced the rigor of a Gymnasium or a *lycée*. After the turn of the century, Hugo Münsterberg, who taught for several years in the United States, extended this point by claiming that with the growing emphasis on research in American universities "the equality of the sexes must disappear in them."[30]

More common were efforts to explain why what worked in the United States could not be successful in Europe. When Cecil Grant asked whether American coeducation could be adopted in England, he noted "the inevitable objection" was, "But the circumstances are so different!" At the International Congress of Women in London in 1899, Dagmar Hjort questioned the applicability for her native Denmark of the success of coeducation in its Scandinavian neighbors; "I will not speak of America," she continued, "which is so far away, and where it is thought circumstances differ so widely from ours that no conclusions can be drawn from comparisons."[31]

Europeans ultimately could not decide whether coeducation was more a cause or a consequence of the elevated position of women in American society. Almost all were certain, however, that mixing the sexes in their own secondary schools would produce much more "immorality" than in the United States. Laboulaye suggested Americans must be "angels," while the Belgian Laveleye informed an American correspondent, "Our morals are very inferior to yours." Godts insisted that in Europe studying with females would not moralize males, but that instead "the stronger sex will corrupt the weaker."[32]

Perhaps the most interesting rejection of American models of both secondary coeducation and degree-granting women's colleges came from leading women educators in England. In the 1890s, Burstall, Zimmern, and three others came to the United States with support from the Gilchrist Trust. They found much to like in public high schools with mixed staffs and witnessed the great value of degrees from the top women's colleges. Yet in 1897 over 150 headmistresses opposed a proposal that Royal Holloway attempt to grant degrees, claiming, "A Women's University means an uncertain and probably a low standard." Eight years later the Association of Headmistresses voted by 152 to 35 to reject coeducation in favor of single-sex secondary schools beginning at age ten. Fears of undercutting the campaign for degrees for women at Oxford and Cambridge and of having men monopolize the headships of coeducational secondary schools led to knowledge of contrary American practices being quietly forgotten.[33]

Conclusion

In terms of direct imitation, neither America's coeducational public high schools nor its degree-granting women's colleges served as models for European reforms in the half century before the First World War. Yet, as the preceding pages have shown, American practices loomed large for both reformers and defenders of the status quo.

188 JAMES C. ALBISETTI

The prominence of women teachers, the mixing of teenage boys and girls in class-rooms, and the generous philanthropy supporting women's colleges suggested possibilities unknown or at least unfamiliar in much of the Old World. Explaining why American practices could not be adopted in their own countries forced many Europeans to talk openly about aspects of their cultures, particularly expectations about gender roles, which might otherwise have been less thoroughly examined.

Notes

1. Carol Strauss Sotiropoulos, *Early Feminists and the Education Debates: England, France, Germany, 1760–1810* (Madison and Teaneck, NJ: Fairleigh Dickinson University Press, 2007); Christina de Bellaigue, *Educating Women: Schooling and Identity in England and France, 1800–1867* (Oxford: Oxford University Press, 2007); Rebecca Rogers, *From the Salon to the Schoolroom: Educating Bourgeois Girls in Nineteenth-Century France* (University Park: Pennsylvania State University Press, 2005), 243–252.

2. Earl Beck, "The German Discovery of American Education," *History of Education Quarterly* 5 (1965): 3–14, here 7; James F. Tent, *Mission on the Rhine: Reeducation and Denazification in American-Occupied Germany* (Chicago: University of Chicago Press, 1982).

3. P.N. Farrar, "American Influence on the Movement for a National System of Elementary Education in England and Wales, 1830–1870," *British Journal of Educational Studies* 14 (1965): 36–47; W.H.G. Armytage, *The American Influence on English Education* (London: Routledge & Kegan Paul, 1967), 17–18; Richard L. Rapson, *Britons View America: Travel Commentary, 1860–1935* (Seattle: University of Washington Press, 1971), 76–78; George Combe, *Notes on the United States of North America during a Phrenological Visit in 1838–9-40* (3 vols.; Edinburgh: Machlachlan Stewart, 1841).

4. Enrico Mayer, "Pubblica educazione negli Stati Uniti d'America," *Antologia* 45, 1 (January 1832): 3–19; Per Adam Siljeström, *Educational Institutions of the United States*, trans. Frederica Brown (New York: Arno Press, 1969 [1853]). 300–309.

5. Combe, *Notes*, 2: 42; Frederick von Raumer, *America and the American People*, trans. William W. Turner (New York: J. and H.G. Langley, 1846), 280; Richard Cobden, *The American Diaries of Richard Cobden*, ed. Elizabeth Hoon Cawley (Princeton: Princeton UP, 1952), 156–157, 162; Siljeström, *Educational Institutions*, 302; Rudolf Dulon, *Aus Amerika: Über Schule, deutsche Schule, amerikanische Schule, und deutsch-amerikanische Schule* (Leipzig and Heidelberg: C.F. Winter, 1866), 251; Emile de Laveleye, "De l'instruction du people au dix-neuvième siècle: L'enseignement populaire dans les écoles américaines," *Revue des deux mondes* 60 (1865): 273–300, here 285.

6. Frances Trollope, *Domestic Manners of the Americans*, ed. Donald Smalley (New York: Knopf, 1949 [1832]), 82, 301; Combe, *Notes*, 3: 242; Frederika Bremer, *The Homes of the New World*, trans. Mary Howitt (2 vols.; New York: Harper & Brothers, 1853), 1: 91, 255–256.

7. Analyses of individual countries include Rapson, *Britons View America*; Jacques Portes, *Une fascination réticente: les États-Unis dans l'opinion française, 1871–1914* (Nancy: Presses universitaires de Nancy, 1990); Delaye Gager, *French Comment on American Education* (New York: Columbia University Press, 1925); Alexander Schmidt, *Reise in die Moderne: Der Amerika-Diskurs des deutschen Bürgertums vor dem ersten Weltkrieg im europäischen Vergleich* (Berlin: Akademie Verlag, 1997); and Andrew J. Torrielli, *Italian Opinion on America as Revealed by Italian Travelers, 1850–1890* (Cambridge, MA.: Harvard University Press, 1941).

8. Wilhelm von Polenz, *Das Land der Zukunft* (Berlin: Fontane, 1903).

9. James Fraser, *Report to the Commissioners Appointed by Her Majesty…on the Common School System of the United States and of the Provinces of Upper and Lower Canada* (London: Eyre & Spottiswoode, 1866), iii–iv; Sophia Jex-Blake, *A Visit to Some American Schools and Colleges* (London: Macmillan, 1867).

10. Édouard Laboulaye, *Paris in America*, trans. Mary L. Booth (New York; Scribners, 1867 [1863]); M.C.M. Simpson, *Letters and Recollections of Julius and Mary Mohl* (London: Kegan Paul, Trench, 1887), 241.

11. Célestin Hippeau, "L'éducation des femmes et des affranchis en Amerique," *Revue des deux mondes* 93 (1869): 450–476, here 461; *L'instruction publique aux États-Unis* (2nd ed.; Paris: Didier, 1872), esp. 96–119.

12. Marino Raichich, "Verso la cultura superiore e le professioni," in *Le donne a scuola: L'educazione femminile nell'Italia dell'Ottocento*, ed. Ilaria Porciani (Florence: Il sedicesimo, 1987), 198; Edoardo Fusco, "L'istruzione femminile di secondo grado," *Il progresso educativo* 1, 6 (August 1869), reprinted in Edoardo Fusco, *Della vita e delle opere di Edoardo Fusco*, ed. his wife (2 vols.; Naples: Tipografia Italiana, 1880–1881), 2: 225–233, here 2: 227; Aristide Gabelli, "L'Italia e l'istruzione femminile," *Nuova Antologia* 15 (1870): 146–167.

13. Dulon, *Aus Amerika*, 171–179; Gotthold Kreyenburg, *Mädchenerziehung und Frauenleben im Aus- und Inlande* (Berlin: Guttentag, 1872), 61–98; John Howard Raymond, *Vassar College, A College for Women in Poughkeepsie, New York* (New York: S.W. Green, 1873); Franz Migerka, *Das Unterrichtswesen in den Vereinigten Staaten* (Vienna: Faesy & Frick, 1877).

14. For example, John Leng, *America in 1876: Pencillings during a Tour in the Centennial Year* (Dundee: Dundee Advertiser, 1877); Ferdinand Buisson, ed., *Rapport sur l'instruction primaire à l'exposition universelle de Philadelphia en 1876* (Paris: Imprimerie Nationale, 1878); Anna Schepeler-Lette, "Einrichtungen für weibliches Erziehungswesen und Frauenerwerb in den Vereinigten Staaten," *Der Arbeiterfreund* 14 (1876): 323–333.

15. William J. Reese, *The Origins of the American High School* (New Haven and London: Yale UP, 1995), 208; Royal Commission on Secondary Education, *Report of the Commissioners* (7 vols.; London: Eyre & Spottiswoode, 1895), 2: 171.

16. Buisson, *Rapport*, 491; Joshua G. Fitch, *Notes on American Schools and Training Colleges* (London: Macmillan, 1890), 27; *Reports of the Mosely Commission to the United States, October–December 1903* (London: Cooperative Printing Society, 1904), 13, 143–144, 165; Carl Grundscheid, *Coeducation in den Vereinigten Staaten von Nord Amerika* (Greifswald: Permetter, 1906), 31; L.R. Klemm, "Das Schulwesen in den Vereinigten Staaten," in *Amerika: Der heutige Standpunkt der Kultur in den Vereinigten Staaten*, ed. Armin Tenner (Berlin: Stuhr, 1886), 33–82, here 53, 61–62.

17. Dulon, *Aus Amerika*, 179; Johann Gottlob Pfleiderer, *Amerikanische Reisebilder* (Bonn: Schergens, 1882), 31; Gabriel Compayré, *L'enseignement secondaire aux États-Unis* (Paris: Hachette, 1896), 2; Sara Burstall, *Impressions of American Education in 1908* (London: Longmans, Green, 1909), 15.

18. Gabriel Compayré, *L'enseignement supérieur aux États-Unis* (Paris: Hachette, 1896), 54; Siljeström, *Educational Institutions*, 311; Jex-Blake, *A Visit*, 26–27; Edward A. Freeman, *Some Impressions of the United States* (London: Longmans, Green, 1883), 178.

19. See James C. Albisetti, "American Women's Colleges through European Eyes, 1865–1914," *History of Education Quarterly* 32, 4 (1992): 439–458; and "European Perceptions of American Coeducation, 1865–1914: Ethnicity, religion, and culture," *Paedagogica Historica* 37, 1 (2001): 123–138.

20. "Vassar Female College," *English Woman's Journal* 9 (August 1862): 401–406; "American University for Women," ibid. 12 (September 1863): 43–47; Hippeau, "L'éducation des femmes," 457–458; Philipp von Nathusius, *Zur Frauenfrage* (Halle: Mühlmann, 1871), 44.

21. Kreyenburg, *Mädchenerziehung*, 95; James Orton, "Four Years in Vassar," *Victoria Magazine* 24 (November 1874): 54–66, here 54; Eduard Boos-Jegher, *Die Tätigkeit der Frau in Amerika* (Bern: Michel & Büchler, 1894), 16; Charles Barneaud, *Origines et progrès de l'éducation en Amérique* (Paris: Savaète, 1898), 343; Sara Burstall, *The Education of Girls in the United States* (London: Sonnenschein, 1894), 112.

22. Ludwig Fulda, *Amerikanische Eindrücke* (Stuttgart and Berlin: Cotta, 1906), 114; Carlo Gardini, *Gli Stati Uniti: Ricordi* (2 vols.; Bologna: Zanichetti, 1891), 1: 161; Migerka, *Das Unterrichtswesen*, 19.

23. Emil Hausknecht, *Amerikanisches Bildungswesen* (Berlin: Gärtner, 1894), 7; E. Stropeno, "Un collège de jeunes filles aux États-Unis: Wellesley College," *Revue internationale de*

l'enseignement 23 (1892): 28–47; G. Bonet-Maury, "Une visite aux collèges de l'enseignement supérieur des jeunes filles aux États-Unis," ibid. 27 (1894): 1–12.

24. Richard Stites, *The Women's Liberation Movement in Russia: Feminism, Nihilism, and Bolshevism, 1860–1930* (Princeton: Princeton University Press, 1978), 75; Delsey cited in Nelleke Bakker and Mineke van Essen, "No Matter of Principle: The Unproblematic Character of Co-education in Girls' Secondary Schooling in the Netherlands, ca. 1870–1930," *History of Education Quarterly* 39, 4 (1999): 460; Maria Grey, *On the Education of Women* (London: Ridgway, 1871), 34–35; idem, "Men and Women: A Sequel," *Fortnightly Review* 29 (1881): 793; Margret Friedrich, *"Ein Paradies ist uns verschlossen…": Zur Geschichte der schulischen Mädchenerziehung in Österreich im "langen" 19. Jahrhundert* (Vienna: Böhlau, 1999), 92.

25. Yvonne Turin, *L'éducation et l'école en Espagne de 1874 à 1902: Libéralisme et tradition* (Paris: Presses universitaires de France, 1959), 211; Caroline Bingham, "'Doing Something for Women': Matthew Vassar & Thomas Holloway," *History Today* 36, 6 (June 1986): 46–51.

26. Patrick K. Bidelman, *Pariahs Stand Up! The Founding of the Liberal Feminist Movement in France, 1858–1879* (Westport, CT: Greenwood, 1982), 106; Camille Sée, ed., *Lycées et collèges de jeunes filles* (6th ed.; Paris: Cerf, 1896), 67–73, 181, 201. See also James C. Albisetti, "The French *Lycées de jeunes filles* in International Perspective, 1878–1910," *Paedagogica Historica* 40, 1/2 (2004): 143–156.

27. Helene Lange, *Lebenserinnerungen* (Berlin: Herbig, 1930), 131.

28. Royal Commission on Secondary Education, *Report*, 5: 563–585; Alice Zimmern, "Notes on the American School System," ibid., 5: 586–590; J.J. Findlay, "Report on Certain Features of Secondary Education in the United States of America, and in Canada," ibid., 7: 333–409; Deirdre Raftery and Susan M. Parkes, *Female Education in Ireland, 1700–1900: Minerva or Madonna* (Dublin; Irish Academic Press, 2007), 119; John Dunstan, "Coeducation and Revolution: Responses to Mixed Schooling in Early Twentieth-Century Russia," *History of Education* 26, 4 (1997): 381.

29. Fraser, *Common School System*, 195; James C. Albisetti, "Catholics and Coeducation: Rhetoric and Reality before *Divini Illius Magistri*," *Paedagogica Historica* 35 (1999): 667–696; Francis Xavier Godts, *Erreurs et crimes en fait d'éducation; le féminisme condamné par les principes de théologie et de philosophie* (Roulers: De Meester, 1903), 237–238, 255.

30. Hugo Münsterberg, *American Traits from the Point of View of a German* (Boston: Houghton Mifflin, 1901), 163; Polenz, *Land der Zukunft*, 236.

31. Cecil Grant, "Can American Coeducation be Grafted upon the English Public School System?" in *Education in the United States*, part 2 [Great Britain Board of Education, *Special Reports on Education Subjects*, vol. 11] (London: H.M. Printing Office, 1902), 85–100, here 85; International Congress of Women, *The International Congress of Women of 1899* (London: Unwin, 1900), 166.

32. Albisetti, "European Perceptions," 130–133; Laboulaye, *Paris in America*, 285; letter from Laveleye in Theodore Stanton, ed., *The Woman Question in Europe* (New York; G.P. Putnam's Sons, 1884), 373n; Godts, *Erreurs et crimes*, 246.

33. James C. Albisetti, "Un-Learned Lessons from the New World? English Views of American Coeducation and Women's Colleges, c. 1865–1910," *History of Education* 29, 5 (2000): 473, 481–483.

Chapter 13

Crossing Borders in Girls' Secondary Education

Joyce Goodman and Rebecca Rogers

This chapter moves beyond the national framework to consider the ways discussions about girls' education crossed European borders. The preceding chapters emphasize how specific national political and cultural contexts shaped the emergence of girls' schools. But, from the outset, foreign ideas and teachers marked the debates and institutional realizations in different countries. Here we explore both the travelling teachers who participated in girls' education outside their home countries, as well as the increasingly international discussions about what girls' secondary schooling should be.

The extra-national or transnational context in which girls' education evolved has received relatively little scholarly attention until recently. Contemporary discourses about globalization, however, have encouraged historians to look at the circulation of ideas, people, and organizations. With respect to girls' secondary education, a few studies have begun to illustrate how such perspectives raise new questions. In the early nineteenth century, Christina de Bellaigue has studied the movement of teachers and students between England and France, while Kay Whitehead has brought to attention the influence of travelling Australian teachers in Great Britain.[1] Historians have also explored how institutions, associations, or networks generated debate and contributed to sharing information about girls' education: the creation of the French *lycée de jeunes filles* (1880) and the curiosity it generated offers one example, as James Albisetti's comparative analysis has shown.[2] Moving forward in time, Joyce Goodman's current research focuses on the range of organizations that underlay international networks among women educationists in the interwar period in particular.[3] These studies allow us to see the interconnections that existed despite the existence of national models.

Clearly, however, there were national models, and another vein of transnational scholarship emphasizes how they circulated, served to inspire, or to repel.[4]

Chapter 12 has already shown the pervasive influence after 1865 of an American model of secondary girls' education for European observers, but this was not the only existing model. Rebecca Rogers has argued that a French *girls' school* model for elites circulated as well, not only in the French colonies but more broadly to England, Canada, and the United States, thanks, in particular, to the role of French female teaching orders.[5] More scholarship exists, however, on the ways an English model of girls' education was exported to the British colonies, alongside the English examination system. This reflects not just the importance of Empire for the understanding of British society but the weight of scholarship on education and empire compared to other European countries. This chapter, while recognizing the role of national models in transnational discussions about girls' education, shifts focus from the models themselves to the individuals, associations, and networks that brought people together. In their discussions, visions of what constituted an appropriate secondary education for girls were far from uniform, revealing the weight of historic circumstances and the complexity of gender relations in an increasingly global world.

Travelling Teachers

Male European cultural elites tended to live cosmopolitan lives, which began during their student days. In the early modern period, young educated elites and university students typically participated in a "grand tour," discovering the art, culture, and society of other European countries. This form of mobility that encouraged cultural and educational exchange was not limited, however, to men, even if women's mobility is less well-known and was probably less extensive. Forms of transnational student exchange that included girls existed in Switzerland, Germany, and eastern France prior to the twentieth-century development of high school study abroad, but scholarship on this subject remains very piecemeal.[6] This section, as a result, will explore the variety of travelling women teachers who have begun to attract attention, ranging from the familiar figure of the governess, to the more mysterious nun, or the underrated lay female missionary.

Seeing the Sights: Governesses and Lady Teachers

Histories of governesses have long focused on the gender and class dynamics of home teaching, ignoring how national considerations often played into this dynamic. Of course, most governesses lived and taught in their own countries, but a significant number did not. For middle- and upper-class families throughout Europe, employing a foreign governess conferred prestige; for the girls within such families, these foreign lady teachers offered a hodgepodge of classes with their native language at the core. For the governesses, of course, experiences teaching abroad introduced them to other cultures and other familial configurations. Whether these experiences had more than an individual impact, however, is up to debate. Still, by the end of the

ANCIENNE INSTITUTION NEYMARK

Penfionat für junge ifraelitifche Mädchen

BOARDING SCHOOL FOR YOUNG LADIES

PENSIONNAT DE MME KAHN

22, Rue Boileau. — AUTEUIL-PARIS (près du bois de Boulogne)

RÉCOMPENSES
De la Société pour l'Instruction élémentaire, et de M. le Ministre de l'Instruction publique

MÉDAILLE D'HONNEUR EN 1877

La situation de la Maison réunit tous les avantages de la ville et de la campagne : l'Institution est à la porte du bois de Boulogne. — Les *Dortoirs*, les *Salles d'étude*, le *Jardin pour les récréations*, et toutes les dépendances sont spécialement et commodément organisés. — Enfin, tout est prévu pour assurer le bien-être des enfants, et pour ajouter notablement à la réputation dont l'Établissement jouit depuis **plus de cinquante ans.**

Figure 13.1   Advertisement for Mme Kahn's Jewish Boarding School in Paris (ca.1880).
Source: Archives de Paris, DT Supplement, n°120, dossier Isaac.

nineteenth century, a variety of formal and informal means existed for teachers to travel abroad, and the development of professional newsletters meant that "seeing the sights" was increasingly translated into prose with probable influence within the classroom.

The travelling experiences of British governesses or lady teachers are better known than of other nationalities, in part because a scholarly tradition existed from at least the early nineteenth century for girls to "finish" their education abroad. As a result, young women travelled to Belgium, Germany, and especially France to study in finishing schools. While learning the language, music, or painting of these countries, aspiring teachers often taught as well, although there were more candidates than positions available. In 1866 a guide for English schoolmistresses warned of an estimated 400 English teachers in Paris, only one-quarter of whom were able to find work.[7] Still, most aspiring teachers used their experiences abroad not just to improve their linguistic abilities and acquire a superficial knowledge of another country's educational system and values, but also because this knowledge translated into higher pay once back home. The Brontë sisters saw this experience abroad as an essential aspect for succeeding in a competitive English educational market: "They say...that without some such step towards attaining superiority we shall probably have a very hard struggle and fail in the end."[8]

The sort of advertisement shown in figure 13.1 above shows how French headmistresses sought to attract both a French and foreign clientele for their schools in the final decades of the nineteenth century

British women were not the only women to cross borders, seeking an educational complement or a means of living. *Crockford's Scholastic Directory* for 1861 signalled, for example, the presence of 489 French governesses in Britain.[9] On the continent, such mobility existed as well, but emerges more randomly in memoirs, private journals, and correspondence. The Alsatian Amélie Weiler, for example, went off to become the governess for the children of an ex-minister of the King of Prussia in 1858 because, at age twenty-six, she had failed to marry, while the German Bertha Buchwald experienced life as a governess in Denmark and Chile after her father fell ill in 1841.[10] Teaching offered single educated women the means to travel and live in other countries, and to discover different political systems and different gender regimes. Their presence in foreign families undoubtedly contributed to forms of cosmopolitanism, about which we as yet know relatively little.[11]

Bellaigue argues in her study of French and English mobility in the first half of the nineteenth century that for teachers and pupils these experiences across the channel resulted in relatively little importing of pedagogical practices and ideas. For the English, in particular, educational travel only reinforced prejudices against French pedagogy as too mechanistic, rigid, and public. Still the importance of this experience reflects women teachers' increasing need for qualifications to teach, and English teachers were influenced by the French emphasis on training, examination, and certification.[12]

Improved travel, the growth of empire, the consolidation of nation-states, and rising educational needs all contributed by the end of the nineteenth century to a heightened mobility among "lady" teachers, encouraged in part by the weight of foreign languages and culture in girls' secondary education. The English were in the vanguard, offering an impressive range of possibilities to discover other educational systems and cultures. In particular, the Gilchrist Travelling Scholarships, founded in 1865, enabled teachers to go to America to investigate the state of schooling and resulted in publications on American education and some pedagogical reform. As head of Manchester High School, Sara Burstall imported American innovations when she introduced courses in household science and typewriting. The British Association of Head Mistresses also organized schoolgirl tours to foreign countries in addition to teacher exchanges.[13] In the process, they constructed visions of empire for girls and developed notions of a suitable curriculum for different groups of students.

In other European countries, the organization of such transnational schooling or teaching opportunities was more haphazard, but they were also on the rise in the twentieth century. Travel literature, in particular, has begun to reveal the importance of this phenomenon in Europe.[14] Organized scholarships requiring reports upon their completion offer some insights into the cultural effects of such educational tourism. In 1905 the French philanthropist Albert Kahn (1860–1940) opened his Around the World fellowships to women *lycée* teachers in France or Algeria; scholarship holders were expected to discover firsthand the situation of girls' education in other countries and broaden their horizons as individuals and as teachers. Among women who visited the United States, some returned urging the development of stronger studies in the sciences, but, on the whole, their observations dealt more with the weight of American women in society than in the nature of studies that might be developed in France.[15] These experiences of short-term mobility undoubtedly affected the individuals involved while rarely calling into question the principles of the educational system within which they operated. For women teachers who settled in other countries, however, the issues were rather different.

Nuns and Women Missionaries

The dictates of a "civilizing mission" had a tremendous impact on women within religious orders as well as within missionary societies, and, for many, this meant setting up schools throughout the world. The Ursuline nun Marie de l'Incarnation is among the most famous of the women religious who left Europe to set up schools for Native Americans in the Canadian hinterlands. Most of the schools set up for "culturally deprived" or "ignorant" native girls were not secondary institutions, however, but some religious orders and a few missionary teachers focused more specifically on elites, opening boarding schools and day schools that reflected the educational convictions of their home country.

Just as British lady teachers travelled the world, visiting and often establishing schools in Australia, New Zealand, or Canada, French nuns similarly ranged widely, opening schools all over the map. By 1890, the teachers within France's largest female missionary order, the Congrégation de Saint-Joseph de Cluny, had set up schools throughout the French colonies, but also in Italy, Portugal, Trinidad, Sierra Leone, Fiji, Peru, and the United States.[16] Most of these schools were simple affairs, but within capital cities the order frequently opened a boarding school for the local or French elites where the tuition received for offering a secondary and "French" education paid for their school for the poor. Alongside specifically missionary orders, the main teaching orders in France also opened schools outside the metropole, particularly after anticlerical laws made teaching difficult and then illegal (1904).

In secondary education, French nuns acquired a reputation for the polished and disciplined training they dispensed within relatively expensive boarding schools or day schools. The Ursulines as well as the elitist Ladies of the Sacred Heart were among the most prominent of these French teaching orders whose "convent schools" attracted pupils in a wide range of colonial and foreign settings. The success of a "French model" of girls' education among elite families is perhaps most striking in English-speaking and predominantly Protestant areas, such as England, Canada, Australia, and the United States.[17] By 1887 in England, thirty-two out of sixty-two apostolic congregations were French in origin, and five were Belgian; together they ran some ninety-five boarding schools. A similar French influence existed in girls' education in Canada: between 1639 and 1939 over 60 percent of the 103 orders of women religious were of French origin.[18] Tellingly, the native Canadian order of Notre-Dame adapted their curriculum to resemble that of the Sacred Heart.[19]

Travelling nuns made an impact not only on pupils but also on future teachers, thanks to the foundation of teacher training institutes in foreign lands. In England, the Belgian order Notre-Dame de Namur opened a prominent teacher training college in Liverpool in 1858.[20] Irish nuns were influential as teachers in the English-speaking world, while both French and German nuns acquired reputations as the educators of the elite in Brazil.[21] Although little studied for the moment, these experiences setting up boarding schools in foreign lands also contributed to the circulation of information about other educational systems, thanks to the universal presence of congregational "newsletters."

Female missionaries provide a final example of the ways travelling women teachers had an impact on girls' education in colonial contexts. Recent scholarship has drawn attention to the role of these women in the early nineteenth century, but clearly their influence increased in the second half of the century as Protestant missionary societies and Jewish associations more openly accepted the activities of "professional" women missionaries. Similarly to nuns, these women were active creating poor schools across the world, but in certain locations where a recognizable elite existed, missionaries offered more advanced schooling for girls. This was the case in Madagascar prior to French colonization in 1896. Here Catholic and Protestant missionaries were in competition and adopted educational strategies targeting specific social groups. The Sisters of Saint-Joseph de Cluny arrived earlier in 1861 and set up schools for poor girls, whereas the Lutheran Norwegian missionaries focused more on the daughters of the elite. In particular, Doctor Borchgrevink and his wife set up a boarding school that thrived from 1872 until 1912, attracting girls from the local aristocracy. More generally, missionary schools forged within Malagasy culture new representations of femininity, where educated women learned to sew and adopted European dress. It would take many more years of the French presence in Madagascar, however, for Malagasy girls to aspire to secondary degrees and "masculine" professions. Only in 1938 did an indigenous girl obtain a scientific baccalaureate; in this society girls aspiring to higher education were mocked by nicknames such as "singing chickens, acting like roosters."[22]

Women missionaries were more influential in establishing secondary institutions for local elites in the Middle East where a demand for advanced schooling was far more developed. As in Madagascar, the competition between different movements and different religions contributed to the establishment of a relatively dense network of institutions. In Palestine, for example, German, French, and even Russian teachers set up boarding schools for girls, although over time the British were the most successful via the Anglican Jerusalem and East Mission schools. Their explicit goal was to introduce the ideal of the Western modern woman to the "good" Christian, Jewish, and Muslim families in the area. This ideal that developed within British secondary institutions took institutional shape during the British Mandate period (1918–1948) through the creation of the Jerusalem Girls' High School and the English High School for Girls in Haifa, where Muslims, Christian, and Jewish girls mingled.[23]

Organizing Beyond the Nation-State

Networks around the Jerusalem Girls High School illustrate flows of information on girls' secondary education through transnational networks that increasingly linked international women's organizations as the twentieth century progressed. The Jerusalem High School provided a venue for meetings of the Israel (Palestine) Association of University Women, founded in 1932 by Miss Hanbidge, formerly headmistress of the Central Foundation School for Girls in London. By 1933/34 the Association had attracted 130 Christian, Jewish, and Muslim members and was working to provide lists of recommended books and of translations into Arabic for the mission's secondary schools in Palestine. Women members also compiled *Education of Women and Girls in Palestine During the Past Hundred Years*, prepared at the request of the American Association of University Women for the "Century of Progress" international exhibition held in Chicago in 1933–1934.[24]

Ideas and practices about girls' education circulating across national borders were transformed as they interacted with political discourses, nation-building, and a variety of approaches to the "woman question." As the following section illustrates, sharing information at times produced common objectives within international women's and teachers' organizations, but most often it revealed the diversity of understandings of femininity within different nation-states, making consensus about girls' education complex.

International Women's Organizations

The major international women's organizations that grew from the 1880s regarded education as a key to the status of women and a basis from which to foster world peace.[25] They saw dissemination of comparative data on girls' schooling as an impetus to change. The International Council of Women published *National Systems of Education* in 1911.[26] The more comprehensive *L'Enseignement secondaire des jeunes filles en Europe* by Amélie Arato, an Hungarian teacher from Budapest, was published under the auspices of the International Federation of University Women (IFUW) in 1934.[27]

Some delegates to international women's organizations articulated "progress" in girls' education in terms of nation-building. At the 1922 IFUW conference, Dr. Emma Formanova-Novokova of the Czecho-Slovakian Federation of University Women, linked freedom from Austrian rule and the establishment of the new Czech Republic with a new and "progressive" situation that had impacted positively on girls' secondary education. At the 1924 IFUW conference her Federation colleague, Dr. Polivoka, spoke of girls' education having progressed to the point where more than half of the 26,500 girls in secondary education in the new republic went on to advanced study; she also portrayed the work and remuneration of men and women teachers having reached a position of parity. For Polivoka, the education of girls and the status of women teachers constituted indices of progress for women and for the new Czech Republic alike.[28]

Other delegates deployed a range of maternalist and equal-rights rhetoric in discussing girls' education and the position of women teachers. Dr. Isabella Grassi, president of the Italian Federation of University Women, told the 1922 IFUW conference that the Italian legislation threatened coeducation in secondary schools and the right of women to teach in the higher boys' or coeducational classes in the middle schools. She called on women to play their part in a "truly maternal spirit of social life." Her spiritual maternalist language was characteristic of the Women's International League for Peace and Freedom, of which she was also a prominent member,[29] and was at variance with the natalist maternalism that underpinned Italian fascism.[30]

In contrast, Germaine Hannevaart, chair of the IFUW Committee on the Interchange of Information on Secondary Education, drew on equal-rights rhetoric in the campaign to resist moves to dismiss married women teachers in Belgium. A founding member of the Open Door International for the Economic Emancipation of the Woman Worker, Hannevart worked to eliminate all gender-specific measures, including for maternity, but nonetheless defended single-sex education for girls. Based on her own experience as a teacher in Brussels, she argued at the IFUW conference in 1924 for separate schools on the grounds that, between the ages of twelve and eighteen, girls' needs differed from those of boys' and because the aim of secondary schooling should not be merely to teach but also to develop character. Still, this schooling should prepare girls to pursue their studies and so she suggested girls' institutions should provide a more systematic preparation for the University for pupils between the ages of fifteen and eighteen.[31]

Like Hannevart, women teachers focussed more on the organization and practice of girls' schooling in their professional discussions in international teachers' organizations than on its broader political or ideological stakes. Here, debate took account of the psychological perspectives that increasingly characterized educational discourse during the interwar period.

International Teachers' Organizations

The women teachers who participated in international conferences expressed a range of opinions about girls' secondary education that cannot be categorized easily as "progressive," "feminist," or "egalitarian." This is strikingly apparent in a conference organized by the Bureau international des fédérations nationales du personnel de l'enseignement secondaire public in The Hague in 1929 around the theme of single-sex schooling versus coeducation in secondary education. The debates illustrate how professional discourse around psychological differences between the sexes and social Darwinistic and eugenic ideas intertwined with national and political narratives and approaches to equality and difference when it came to defining objectives and practice in girls' secondary education.[32]

There was almost unanimous opposition from the women contributors to the views of Zofia Degen-Slosarska, head of the girls' secondary school in Warsaw, Poland. Her social Darwinistic portrayal of the onward march of civilization, differentiating the sexes and classes, led her to argue for distinctly different schooling for boys and girls. An admirer of Gina Lombroso, the Italian sociologist and historian, whose coauthored work on criminology with her father, Cesare Lombroso, was highly influenced by approaches from physiognomy and social Darwinism, Degen-Slosarska believed that women's desire for political and social liberty, or their affirmation of the right to higher education, was detrimental both to the nation and women themselves. She used Gina Lombroso's insights on the psychic traits of little girls and boys to criticize changes to Polish primary and secondary schooling since 1916, which, with the exception of some small differences in manual training and gymnastics, now provided the same program of maths, physics, and Latin to both sexes. She highlighted how identical programs in the higher schools lacked child psychology, domestic economy, manual training, and general psychology, and argued that, like coeducation, these programs did not satisfy girls' needs for more concrete methods of teaching.

While some participants shared Degen-Slosarska's social Darwinism in varying degrees, few were so hostile to advanced study for women. The Belgian Dr. van der Noot, for example, drawing on ideas found in the work of American G. Stanley Hall, similarly warned against the dangers of "masculinizing", notably for girls between the ages of twelve and fifteen, when, she argued, the feminine organism was particularly fragile and prone to anemia, chlorosis, scoliosis, and the onset of tuberculosis. But once a girl's organism had reached a level of stability around the age of fifteen and her personality was formed, she could undertake intense and theoretical study like that of boys, and it was logical that girls should enjoy the same rights of entry into higher education. (Van der Noot herself held a doctorate in physical chemistry.) Indeed, she argued, young Belgian women educated under feminine direction had proved that an intellectual elite could be formed without engendering what she termed an "excessive feminism, eccentric and disgraceful."

The Bureau's general secretary, the socialist Frenchwoman Suzanne Collette, was the most outspoken in her criticism of the view that girls' schooling should conform to the future role of women in society. She defended a vision of secondary education in which all pupils, irrespective of sex or class, had an equal right to receive the education for which they were capable; collective interest demanded that all intelligences should be cultivated and developed to the full extent possible. Refusing to perceive girls' education as a "problem" requiring specific provisions, she defended systems providing the same conditions of recruitment, programs, hours for academic and cultural courses, examinations, and rights. Collette was not insensitive to prevailing discourses of psychology. This was not, however, a basis for separation of the sexes, but a question of differentiation between advanced and less advanced pupils. She defended a new conception of moral formation for both sexes that took into account a more egalitarian vision of citizenship. Given women's responsibilities in the family, motherhood was particularly pertinent to girls, but it needed to be seen in the context of women's work and service to collective life. A modern course of education for girls should then consist of "scientific" lessons in hygiene and domestic economy, managing a budget, fitting out and decorating the home, and lessons in medicine, maternity, and child psychology, lessons that would help women cope with their double burden. While some women delegates spoke on the basis of their experience, a large number of the male and female participants appeared to have little knowledge of single-sex education, and when it came to voting on resolutions, they passed without debate or challenge.[33]

From 1933 onward, women's discussion in international teachers' organizations increasingly focussed on the effects of world economies and unemployment, the militarization of education, the rise of fascism, and their consequences for women teachers, while international women's organizations paid particular attention to the situation of German women teachers, and to the need for married women to continue to work with full salary. In this political context, coeducation and the psychological development of young women were less urgent questions.[34] As war progressed across Europe and schools were displaced, many teachers' and women's organizations were proscribed and ceased to function.

The postwar emergence of mass secondary education and the rising age of obligatory schooling changed fundamentally the nature of transnational discussions about girls' secondary education. Access to schools and to examinations allowing further study had ceased to be an issue in most European countries, just as coeducation had increasingly become the norm. Increasingly, the associations concerned about girls' education looked beyond Europe. UNESCO placed concern about discriminations in education at the heart of numerous initiatives. Its 2000 World Education Forum at Dakar affirmed a commitment by 2005 to eliminate the gender disparities in secondary education that could still be identified, and by 2015 to achieve gender equality in education, a goal cosmopolitan women educators can all applaud, even if history renders one skeptical.[35]

Notes

1. Christina de Bellaigue, *Educating Women. Schooling and Identity in England and France, 1800–1867* (Oxford: Oxford University Press, 2007); Lynne Trethewey and Kay Whitehead, "Beyond Centre and Periphery: Transnationalism in Two Teacher/suffragettes' Work," *History of Education* 32, 5 (2003): 547–559; Kay Whitehead and Judith Peppard, "Transnational

Innovations, Local Conditions and Disruptive Teachers and Students in Interwar Education," *Paedagogica Historica* 42, 1/2 (2006): 177–189.

2. James C. Albisetti, "The French *Lycées de jeunes filles* in International Perspective, 1878–1910," *Paedagogica Historica* 40, 1/2 (2004): 143–156.

3. Joyce Goodman, "Working for Change Across International Borders: The Association of Headmistresses and Education for International Citizenship," *Paedagogica Historica* 43, 1 (2007): 165–180.

4. Rebecca Rogers, "Questioning National Models: The History of Women Teachers in a Comparative Perspective," paper delivered at International Federation for Research in Women's History conference, Sydney, July 9, 2005, available at *www.historians.ie/women/ rogers.PDF*.

5. Rebecca Rogers, *From the Salon to the Schoolroom: Educating Bourgeois Girls in Nineteenth-Century France* (University Park: Pennsylvania State University Press, 2005), 227–252.

6. Pierre Caspard, "Les changes linguistiques adolescents. Une pratique éducative, XVIIe–XIXe siècles," *Revue Historique Neuchâteloise* 1–2 (2000): 5–85.

7. See Bellaigue, *Educating Women*, 200–230; and Rogers, "French Education for British Girls in the Nineteenth Century," *Women's History Magazine* 42 (2002): 21–29.

8. C. Brontë to E. Branwell, Rawdon, September 29, 1841, cited in Bellaigue, *Educating Women*, 206.

9. Cited in Bellaigue, *Educating Women*, 200, 220.

10. Amélie Weiler, *Journal d'une jeune fille mal dans son siècle, 1840–1859* (Strasbourg: Éditions La Nuée Bleue, 1994); James C. Albisetti, *Schooling German Girls and Women: Secondary and Higher Education in the Nineteenth Century* (Princeton: Princeton University Press, 1988), 74–81.

11. Eric Mension-Rigau, *Aristocrates et grands bourgeois. Education, traditions, valeurs* (Paris: Plon, 1994), 299–338.

12. Bellaigue, *Educating Women*, 215.

13. Joyce Goodman, " 'Their Market Value Must Be Greater For The Experience They Had Gained': Secondary School Headmistresss and Empire, 1897–1914," in *Gender, Colonialism and Education. The Politics of Experience*, ed. Joyce Goodman and Jane Martin (London: Woburn Press, 2002), 175–198.

14. Loukia Efthymiou, "Récits de voyage. Quatre enseignantes à la Belle-Époque," *Clio. Histoire, femmes et sociétés* 28 (2008): 133–144.

15. Whitney Walton, "Des enseignantes en voyage: les rapports des boursières Albert Kahn sur la France et les États-Unis, 1898–1930," in *Le voyage au féminin. Perspective historiques et littéraires (XVIIe–XXe siècles)*, ed. Nicolas Bourguinat (Strasbourg: Presses universitaires de Strasbourg, 2008), 131–149.

16. Private archives of the Congrégation de Saint-Joseph de Cluny, 156 AP I, 207, "Établissements de la congrégation de Saint-Joseph de Cluny dans les colonies françaises et à l'étranger."

17. See the case-study of the schools of the Sacred Heart, Christine Trimingham Jack, *Growing Good Catholic Girls. Education and Convent Life in Australia* (Melbourne: Melbourne University Press, 2003).

18. Elizabeth Smyth, "French Women Religious on the Canadian Frontier: A Case Study of Two Nineteenth-Century Convent Academies of the Sisters of St Joseph (Toronto) and the Religious of the Sacred Heart of Jesus (London)," in *Legacy and Contribution to Canada of European Female Emigrants*, ed. F. LeJeune (Geneva: Peter Lang, 2003), 155–174.

19. Marta Danylewycz, *Taking the Veil: An Alternative to Marriage, Motherhood and Spinsterhood in Quebec, 1840–1920* (Toronto: McClelland and Stewart, 1987), 77.

20. Susan O'Brien, "French Nuns in 19th-century England," *Past and Present* 154 (1997): 142–180; see, as well, Carmen Mangion, *Contested Identities: Catholic Women Religious in Nineteenth-century England and Wales* (Manchester: Manchester University Press, 2008).

21. Angela Xavier de Brito, " 'Le solde est positif.' Culture scolaire et socialisation des élites féminines au Brésil, 1920–1970," *Éducation et sociétés* 15 (April–June 2005): 153–167.

22. Jacqueline Ravelomanana-Randrianjafinimana, *Histoire de l'éducation des jeunes filles malgaches. Du XVIe siècle au milieu du XXe siècle (exemple Merina -Madagascar)* (Imarivolanitra: Éditions Antso, 1996), 134–138, 284–285.

23. Inger Marie Okkenhaug, *The Quality of Heroic Living, of High Endeavor and Adventure: Anglican Mission, Women and Education in Palestine*, 1888–1948 (Leiden, Boston, Cologne: Brill, 2002).

24. Anon, *History of the Israel (Palestine) Association of University Women* (Jerusalem, n.p., n.d.)

25. Christa Kersting, "Weibliche Bildung and Bildungspolitik: das International Council of Women und seine Kongresse in Chicago (1893), London (1899) und Berlin (1904)," *Paedagogica Historica* 44, 3 (2008): 327–346.

26. Mrs. Ogilvie Gordon, *National Systems of Education. First Report of the Education Committee of the International Council of Women* (Aberdeen: Rosemount Press, NCW, 1911), 10–11.

27. Joyce Goodman, "Social Change and Secondary Schooling for Girls in the 'Long 1920s': European Engagements," *History of Education* 36, 1/2 (2007): 497–514.

28. Joyce Goodman, "Cosmopolitan Women Educators, 1920–1939, Inside-outside Activism and Abjection," *Paedagogica Historica* 42, 1 (2010), forthcoming.

29. Goodman, "Cosmopolitan Women Educators."

30. Ann Taylor Allen, *Feminism, and Motherhood in Western Europe 1890–1970* (New York: Palgrave Macmillan, 2005), 60.

31. Goodman, "Cosmopolitan Women Educators."

32. For women's arguments, see "Le Congrès de la Haye, l'enseignement secondaire des jeunes filles," *Bulletin international des federations nationales du personnel de l'enseignement secondaire public* 24 (March 1929), 17–43; and 25 (June 1929), 33– 60, which carry the reports of affiliated national federations and include comments on the points made by some of these women.

33. Goodman, "Social Change."

34. Goodman, "Cosmopolitan Women Educators."

35. http://www.unesco.org.uk/UserFiles/File/EFA_Mapping.pdf (accessed June 16, 2009).

Contributors

James Albisetti is Professor of History at the University of Kentucky. Author of *Secondary School Reform in Imperial Germany* (1983), *Schooling German Girls and Women* (1989), and over thirty articles and chapters on German and comparative educational history, he has served as president of the History of Education Society (USA) and had two terms on the executive committee of the International Standing Conference for the History of Education. He is nearing completion of a new book, *Eminent Immigrant Victorians: The Nineteenth Century of Salis and Julie Schwabe*.

Hilda Amsing is Assistant Professor in Education at the University of Groningen. She completed a Ph.D. on the history of Dutch secondary education. She is currently researching educational innovation after the Second World War. She has edited two books in Dutch on educational quality and has published articles in both national and international journals.

Helena C. Araújo is Professor at the University of Porto (Portugal) in the Faculty of Education. She is a member of the editorial board of the *Journal of Social Science Education & Educação, Sociedade e Culturas* (Education, Society and Cultures) and was Director of *ex-aequo* (Journal Portuguese Association of Women Studies, 2002–07). She has publications in journals in the sociology of education, gender studies, youth and citizenship studies, namely in *Gender and Education*, *International Studies in Sociology of Education, European Journal of Women Studies* and books and chapters in Routledge, Open University, and Afrontamento.

Krassimira Daskalova is Associate Professor in Modern European Cultural History at the Faculty of Philosophy, St. Kliment Ohridski University of Sofia. Her research and teaching interests are in the fields of women's/gender history, history of the book and reading, and Southeast European cultural history. She is coeditor of *Aspasia. International Yearbook of Central, Eastern and Southeastern European Women's and Gender History* and of *L'Homme, Europäische Zeitschrift für Feministische Geschichtswissenschaft*. Krassimira Daskalova is the current president of the International Federation for Research in Women's/Gender History www.ifrwh.com.

E. Thomas Ewing is Associate Professor of History at Virginia Tech, where he teaches courses in European, Middle Eastern, and world history, women's history, and historical methods. His publications include *The Teachers of Stalinism. Policy, Practice, and Power in Soviet Schools in the 1930s* (2002) and articles in *Gender &*

History, American Educational Research Journal, History of Education Quarterly, Russian Review, and *The Journal of Women's History.* He is editor of *Revolution and Pedagogy. Transnational Perspectives on the Social Foundations of Education* (2005) and coeditor of *Education and the Great Depression. Lessons from a Global History* (2006). His current research is on coeducational and single-sex schooling in modern Russia.

Consuelo Flecha is Professor of the History of Women's Education at the University of Seville in Spain. She has published widely in women's history in addition to work in the history of women's education. Her particular research focus is on women's secondary and university education. She is active in women's history circles in Spain and organized the XIIth International Conference of the AEIHM in Seville in 2004 (the Spanish Association for Women's History).

Laura Fonseca is Auxiliary Professor in the Faculty of Education at the University of Porto. Her doctorate is in Sociology of Education. Her research concerns the sociology of education, gender and youth studies. Her book *Justiça Social e Educação* (Social Justice and Education) focuses on young women and educational transitions in the 2000s, exploring how they experience social and ethnic differences.

Joyce Goodman is Professor of History of Education and Dean of the Faculty of Education, Health and Social Care, at the University of Winchester, UK. She is a former editor of the journal, *History of Education,* president of the History of Education Society GB, and former secretary of the International Standing Conference for the History of Education. Her current research focuses on gender, imperialism, internationalism, and education.

Eliane Gubin is Professor in Social History at the Université libre de Bruxelles (ULB) and is a specialist in women's history. In 1989, she founded the first academic group to promote research on women at ULB (GIEF, Groupe interdisciplinaire d'études sur les femmes), and in 1992, she founded the scientific review, *Sextant.* She is codirector of the Centre d'Archives pour l'Histoire des femmes in Brussels since 1995. Coeditor of the *Dictionnaire des femmes belges XIXe-XXe s.* (Ed. Racine, Bruxelles, 2006), she has also published an Encyclopédie *d'histoire des femmes en Belgique* (XIXe–XXe s.). (Ed. Racine, Bruxelles, 2009).

Juliane Jacobi is Professor of History of Education at the Institut für Erziehungwissenschaft (Institute for Educational Sciences) at the Universität Potsdam in Germany. In the course of her career she has been Professor in the History of Education and Gender Studies at the University of Bielefeld as well as a Visiting Professor in Gender Studies at the Central European University-Budapest. She has published important work on the history of girls' education in Germany for over twenty years, including the chapter on education in the German edition of the prestigious *History of Women in the West* (vol. 4).

Agneta Linné is Professor Emeritus of Education at Örebro University, Sweden, and historian of education, of teacher education, and of women. She has headed or been cohead of a number of research projects for the Swedish Research Council and other funding bodies, including *When Practical Knowledge Meets Academia: Continuity and*

Change in Teacher Education and *Shaping the Public Sphere: a Collective Biography of Stockholm Women 1880–1920.* She is the vice chair of the network of historical research of the Nordic Educational Research Association since 2002. Her research has been disseminated in international and national journals.

Cristina Rocha is Associate Professor in the Faculty of Education at the University of Porto (Portugal); her fields of expertise include family and childhood studies, and the socio-history of secondary schooling and certain professions (pharmacy). One of her main publications is *As mulheres e a cidadania, as mulheres e o trabalho na esfera pública e na esfera doméstica* (Women, citizenship and work in public and private spheres, 2006, coauthor).

Rebecca Rogers is Professor in the History of Education at Université Paris Descartes (Paris 5) and member of the research laboratory: l'UMR 8070 Centre de recherches sur le lien social. Specialist in the history of middle-class girls' education in France, she has published in both English and French on this subject as well as on coeducation. As a board member of *Histoire de l'Education* she has directed two special issues on women teachers and girls' education in 2003 and 2007. She is currently preparing a biography on the woman who founded the first school for Muslim girls in Algiers in 1845.

Simonetta Soldani is Professor of Contemporary History at the University of Florence and coordinator of the Doctoral Program in Modern and Contemporary History of the same University. Since 1982 she has been codirector of *Passato e presente*, a journal of contemporary history. In the past twenty years she has published widely in women's and educational history, with a special focus on the dynamics concerning the emergence, in the course of the nineteenth century, of a new model of "Italian Womanhood" and on its relationship with the building of a modern Italian State. More recently, her work has focused on women's access to new professions like teaching or writing for magazines.

Mineke van Essen is a historian of education and Professor Emeritus of Gender Studies and Education at the University of Groningen, The Netherlands. Her main scholarly expertise concerns the position of girls and women in education, and the history of teacher training. On both subjects she has written monographs in Dutch. Also, she has published widely in international and national journals. Currently she is preparing a biography of the first Dutch woman professor in special education.

English-language Bibliography

(Multilingual references can be culled from the footnotes of individual chapters)

Winiarz, Adam. "Girls' Education in the Kingdom of Poland (1815–1915)." In *Women in Polish Society*, ed. Rudolf Jaworski and Bianka Pietrow-Ennker, 91–110. Boulder, CO: East European Monographs, 1992.

Aksit, Elif Ekin. "Girls' Education and the Paradoxes of Modernity and Nationalism in the late Ottoman Empire and the Early Turkish Republic." Ph. D. diss., State University of New York at Binghamton, 2004.

Alberdi, Isabel and Inés Alberdi. "Spain." In *Girls and Young Women in Education: A European Perspective*, ed. Maggie Wilson, 153–169. Oxford: Pergamon Press, 1990.

Albisetti, James C. "American Women's Colleges through European Eyes, 1865–1914." *History of Education Quarterly* 32 (1992): 439–458.

———. "Catholics and Coeducation: Rhetoric and Reality in Europe before *Divini Illius Magistri*." *Paedagogica Historica* 35, 3 (1999): 667–696.

———. "European Perceptions of American Coeducation, 1865–1914: Ethnicity, Religion, and Culture." *Paedagogica Historica* 37, 1 (2001): 123–138.

———. "Female Education in German-Speaking Austria, Germany and Switzerland." In *Austrian Women in the Nineteenth and Twentieth Century*, ed. Davis Good, Margret Grandner, and Mary Jo Maynes, 39–57. Providence and Oxford: Berghan, 1996.

———. "The French *Lycée de jeunes filles* in International Perspective, 1878–1910." *Paedagogica Historica* 40, 1/2 (2004): 143–156.

———. *Schooling German Girls and Women: Secondary and Higher Education in the Nineteenth Century*. Princeton: Princeton University Press, 1988.

———. "Un-Learned Lessons from the New World? English Views of American Coeducation and Women's Colleges, c. 1865–1910." *History of Education* 29, 5 (2000): 473–489.

Aldrich, Richard, ed. *Public or Private Education? Lessons from History*. London: Woburn 2004.

Allen, Ann Taylor. *Feminism and Motherhood in Western Europe, 1890–1970*. New York: Palgrave MacMillan, 2005.

Allender, Tim. "Imagining Innovation: State Agendas for Women's Education in Colonial India." *History of Education Researcher* 80 (November 2007): 100–112.

Alston, Patrick. *Education and the State in Tsarist Russia*. Stanford: Stanford University Press, 1969.

Araújo, Helena C. "The Emergence of a 'New Orthodoxy': Public Debates on Women's Capacities and Education in Portugal (1880–1910)." *Gender and Education—Women's Education in Europe* 4, 1/2 (1992): 7–24.

———. "Mothering and Citizenship. Educational Conflicts in Portugal." In *Challenging Democracy: International Perspectives on Gender and Citizenship*, ed. Madeleine Arnot and Jo-Anne Dillabough, 105–121. London: Routledge/Falmer Press, 2000.

———. "Pathways and Subjectivities of Portuguese Women Teachers through Their Life Histories, 1919–1933." In *Telling Women's Lives*, ed. Kathleen Weiler and Sue Middleton, 113–129. Middlesex: Open University, 1999.

Arnot, Madeleine, Miriam David, and Gaby Wiener. *Closing the Gender Gap: Postwar Education and Social Change*. London: Polity Press, 1999.

Bakker, Nelleke and Mineke van Essen. "No Matter of Principle: The Unproblematic Character of Co-education in Girls' Secondary Schooling in the Netherlands, ca. 1870–1930." *History of Education Quarterly* 39, 4 (1999): 454–475.

Ballarín, Pilar and Cándida Martinez. "Women And Higher Education." In *New Women of Spain. Social-Political and Philosophical Studies of Feminist Thought*, ed. Elisabeth de Sotelo, 429–441. Münster: Lit Verlag, 2005.

Barbagli, Marzio. *Educating for Unemployment: Politics, Labor Markets, and the School System—Italy, 1859–1972.* Trans. Robert H. Ross. New York: Columbia University Press, 1982.

Barker, Hannah and Elaine Chalus, eds. *Women's History: Britain, 1700–1850. An Introduction.* London: UCL Press, 2005.

Bellaigue, Christina de. *Educating Women. Schooling and Identity in England and France 1800–1867.* Oxford: Oxford University Press, 2007.

Bereday, George Z.F. , William W. Brickman, and Gerald H. Read with the assistance of Ina Schlesinger, eds. *The Changing Soviet School: The Comparative Education Society Field Study in the U.S.S.R.* Boston: Houghton Mifflin Company, 1960.

Bhattacharya, Sabyasachi, Joseph Bara, Chinna Roa Yagati, and B.M.Sankhdher, eds. *Development of Women's Education in India, 1850–1920.* New Delhi: Kanishka Publishers, 2001.

Bjurman, Eva Lis. "Sophie, Education and Love: A Young Bourgeois Girl in Denmark in the 1790s." *Ethnologica Scandinavica* 26 (1996): 5–24.

Blom, Ida. "Gender and Nation in International Comparison." In *Gendered Nations: Nationalisms and Gender Order in the Long Nineteenth Century*, ed. Ida Blom, Karen Hagemann, and Catherine Hall, 3–26. Oxford: Berg, 2000.

———. "Nation-Class-Gender: Scandinavia at the Turn of the Century." *Scandinavian Journal of History* 21, 1 (1996): 1–16.

Brooksbank Jones, Anny. *Women in Contemporary Spain.* Manchester: Manchester University Press, 1997.

Clark, Linda L. *The Rise of Professional Women in France: Gender and Public Administration since 1830.* Cambridge: Cambridge University Press, 2000.

———. *Women and Achievement in Nineteenth-Century Europe.* Cambridge: Cambridge University Press, 2008.

Cohen, Michèle. "Gender and 'Method' in Eighteenth-century English Education." *History of Education* 33, 5 (2004): 585–593.

Conway, Jill. "Women Reformers and American Culture, 1870–1930." *Journal of Social History* 5, 2 (1971–1972): 164–177.

Copelman, Dina. *London's Women Teachers: Gender, Class and Feminism, 1870–1930.* London: Routledge, 1996.

Dawtrey, Liz, Janet Holland, Merrill Hammer, with Sue Sheldon, eds. *Equality and Inequality in Education Policy.* Clevedon: Multilingual Matters, 1995.

Deem, Rosemary, ed. *Coeducation Reconsidered.* Milton Keynes: Open University Press, 1984.

Dowler, Wayne. *Classroom and Empire. The Politics of Schooling Russia's Eastern Nationalities, 1860–1917.* Montreal: McGill-Queen's University Press, 2001.

Dunstan, John. "Coeducation and Revolution: Responses to Mixed Schooling in Early Twentieth-Century Russia." *History of Education* 26, 4 (1997): 375–393.

Dyhouse, Carol. *Girls Growing up in Late Victorian and Edwardian England.* London: Routledge and Kegan Paul, 1981.

Eklof, Ben. *Russian Peasant Schools: Officialdom, Village Culture, and Popular Pedagogy, 1864–1914.* Berkeley: University of California Press, 1986.

Enders, Victoria Lorée and Pamela Beth Radcliff, eds. *Constructing Spanish Womanhood: Female Identity In Modern Spain.* New York: State University of New York Press, 1999.

Essen, Mineke van. "Anna Barbara Van Meerten-Schilperoort (1778–1853). Feminist pioneer?" *Revue Belge de Philologie et d'Histoire* 77 (1999): 383–401.

———. "'New' Girls and Traditional Womanhood. Girlhood and Education in the Netherlands in the Nineteenth and Twentieth Century." *Paedagogica Historica* 29, 1 (1993): 125–151.

———. "Strategies of Women Teachers 1860–1920. Feminization in Dutch Elementary and Secondary Schools from a Comparative Perspective." *History of Education* 28, 4 (1999): 413–433.

Ewing, E. Thomas. "The Repudiation of Single-Sex Schooling: Boys' Schools in the Soviet Union, 1943–1954." *American Educational Research Journal* 43, 4 (2006): 621–650.

———. "A Stalinist Celebrity Teacher: Gender and Professional Identities in the Soviet Union in the 1930s." *Journal of Women's History* 16 (Winter 2004): 92–118.

Flecha Garcia, Consuelo. "Women at Spanish Universities." In *New Women of Spain: Social-Political and Philosophical Studies of Feminist Thought*, ed. Elisabeth de Sotelo, 397–409. Münster: Lit. Verlag, 2005.

Fletcher, Sheila. *Feminists and Bureaucrats. A Study in the Development of Girls' Education in the Ninteenth Century.* Cambridge: Cambridge University Press, 1980.

Fraser, Nancy. "Rethinking the Public Sphere." In *Habermas and the Public Sphere*, ed. Craig Calhoun, 109–142. Cambridge, MA: MIT Press, 1992.

Gold, Carol. *Educating Middle-Class Daughters: Private Girls Schools in Copenhagen 1790–1820.* Copenhagen: The Royal Library & Museum Tusculanum Press, 1996.

Goodman, Joyce. "At the Centre of a Circle Whose Circumference Spans All Nations: The Ladies Committee of the British and Foreign School Society, 1813–1837." In *Women, Religion and Feminism in Modern Britain, 1750–1900*, ed. Susan Morgan, 53–70. Houndsmill and New York: Palgrave MacMillan, 2002.

———. "'A Cloistered Ethos?' Landscapes of Learning and English Secondary Schools for Girls: An Historical Perspective." *Paedagogica Historica* 41, 4–5 (2005): 589–603.

———. "Cosmopolitan Women Educators, 1920–1939: Inside/Outside Activism and Abjection." *Paedagogica Historica* 46, 1/2 (2010).

———. "Social Change and Secondary Schooling for Girls in the 'Long 1920s': European Engagements." In *Social Change in the History of British Education*, ed. Joyce Goodman, Gary McCulloch, William Richardson, 95–112. London: Routledge, 2008.

Goodman, Joyce and Jane Martin, eds. *Gender, Colonialism and Education: The Politics of Experience.* London: Woburn, 2002.

Goodman, Joyce and Sylvia Harrop, eds. *Women Educational-Policy Making and Administration in England. Authoritative Women since 1800.* London: Routledge, 2000.

Gouda, Frances. *Dutch Culture Overseas: Colonial Practice in the Netherlands Indies, 1900–1942.* Amsterdam: Amsterdam University Press, 1995.

Grazia. Victoria de. *How Fascism Ruled Women: Italy, 1922–1945.* Berkeley: University of California Press, 1992.

Gruber, Helmut and Pamela Graves, eds. *Women and Socialism, Socialism and Women. Europe Between the Two World Wars.* New York; Oxford: Berghan Books, 1988.

Haan, Francisca de, Krassimira Daskalova, and Anna Loutfi, eds. *A Biographical Dictionary of Women's Movements and Feminisms. Central, Eastern and South Eastern Europe. 19th and 20th Centuries.* Budapest: Central European University Press, 2006.

Hans, Nicholas. *History of Russian Educational Policy (1701–1917).* New York: Russell & Russell, 1964.

Hilton, Mary and Jill Shefrin, eds. *Educating the Child in Enlightenment Britain.* Surrey: Ashgate, 2009.

Holmes, Larry E. *The Kremlin and the Schoolhouse. Reforming Education in Soviet Russia, 1917–1931.* Bloomington: Indiana University Press, 1991.

Hunt, Felicity. *Gender and Policy in English Education 1902–1944.* Brighton: Harvester Wheatsheaf, 1991.

———, ed. *Lessons for Life. The Schooling of Girls and Women, 1850–1950.* Oxford: Blackwell, 1987.

Jacobs, Andrea. "'The Girls Have Done Very Decidedly Better than the Boys'; Girls and Examinations 1860–1902." *Journal of Educational Administration and History* 33, 2 (2001): 120–136.

Johanson, Christine. *Women's Struggle for Higher Education in Russia, 1855–1900.* Kingston, ON: McGill-Queen's University Press, 1987.

Johnson, William H.E. *Russia's Educational Heritage.* Pittsburgh: Carnegie Press, 1950.

Jones, Gareth Elwyn. *Education and Female Emancipation. The Welsh Experience, 1847–1914.* Cardiff: University of Wales Press, 1990.

Knott, Sarah and Barbara Taylor, eds. *Women, Gender and Enlightenment.* Basingstoke: Palgrave, 2005.

Kosambi, Meera. "A Window in the Prison-House: Women's Education and the Politics of Social Reform in Nineteenth-Century Western India." *History of Education* 29, 5 (2000): 429–442.

Linné, Agneta. "Morality, the Child, or Science? A Study of Tradition and Change in the Education of Elementary School Teachers in Sweden." *Journal of Curriculum Studies* 31 (1999): 429–447.

Lougee, Carolyn. "Noblesse, Domesticity, and Social Reform: The Education of Girls by Fénelon and Saint-Cyr." *History of Education Quarterly* 14, 1 (Spring 1974): 87–113.

Lowe, Roy. *Schooling and Social Change, 1964–1990*. London: Routledge, 1997.

Mangion, Carmen. *Contested Identities: Catholic Women Religious in Nineteenth-Century England and Wales*. Manchester: Manchester University Press, 2004.

Margadant, Jo Burr. *Madame le Professeur: Women Educators in the Third Republic*. Princeton: Princeton University Press, 1990.

Mavrinac, Marilyn. "Conflicted Progress : Coeducation and Gender Equity in Twentieth-Century French School Reforms." *Harvard Educational Review* 67, 4 (Winter 1997): 772–795.

Messbarger, Rebecca. *The Century of Women: Representations of Women in Eighteenth-Century Italian Public Discourse*. Toronto: University of Toronto Press, 2002.

Mianda, Gertrude. "Colonialism, Education and Gender Relations in the Belgian Congo: The Évolué Case." In *Women in African Colonial Histories*, ed. Jean Allman, Susan Gerger, and Nakanyike Musisi, 144–163. Bloomington: Indiana University Press, 2002.

Moore, Lindy. "Young Ladies' Institutions: The Development of Secondary Schools for Girls in Scotland, 1833–1870." *History of Education* 32, 3 (2003): 249–272.

Morcillo, Aurora G. *True Catholic Womanhood: Gender Ideology in Franco's Spain*. De Kalb, IL: Northern Illinois University Press, 2000.

Morris Matthews, Kay. *In Their Own Right: Women and Higher Education in New Zealand before 1945*. Wellington: NZCER Press, 2008.

Nash, Carol S. "Educating New Mothers: Women and the Enlightenment in Russia." *History of Education Quarterly* 21, 3 (Autumn, 1981): 301–316.

Nash, Mary. "The Rise of the Women's Movement in Nineteenth-century Spain." In *Women's Emancipation Movements in the 19th Century: A European Perspective*, ed. Sylvia Paletschek and Bianka Pietrow-Ennker, 243–262. Stanford: Stanford University Press, 2004.

———. "Towards a New Moral Order: National Catholicism, Culture and Gender." In *Spanish History since 1808*, ed. José Álvarez Junco and Adrian Shubert, 289–302. London and New York: Arnold and Oxford University Press, 2000.

Nóvoa, António. "The Teaching Profession in Europe." In *Problems and Prospects in European Education*, ed. Elizabeth Sherman Swing, Jürgen Schriewer, and François Orivel, 45–71. Westport: Praeger 2000.

———. "Ways of Saying, Ways of Seeing—Public Images of Teachers (19th–20th Centuries)." *Paedagogica Historica* 36, 1 (2000): 21–43.

Nóvoa, António, Marc Depaepe, and Erwin Johanningmeier, eds. *The Colonial Experience in Education: Historical Issues and Perspectives*. *Paedagogica Historica*, supplementary series, Vol. 1. Gent: Stichting Paedagogica Historica, 1995.

Nyström, Per. *Mary Wollstonecraft's Scandinavian Journey*. Gothenburg: Acta Regiae Societatis Scientiarum et Litterarum Gothoburgensis, 1980.

O'Brien, Susan. "French Nuns in Nineteenth-Century England." *Past and Present* 154 (February 1997): 142–180.

Offen, Karen. "The Second Sex and the *Baccalauréat* in Republican France, 1880–1924." *French Historical Studies* 3, 2 (Fall 1983): 252–288.

Okkenhaug, Inger Marie, ed. *Gender, Race and Religion: Nordic Missions 1860–1940*. Uppsala: Studia Missionalia Svecana XCI, 2003.

———. *The Quality of Heroic Living, of High Endeavor and Adventure: Anglican Mission, Women and Education in Palestine, 1888–1948*. Leiden; Boston; Cologne: Brill, 2002.

Parkes, Susan, ed. *A Danger to the Men: A History of Women in Trinity College Dublin, 1904–2004*. Dublin: The Lilliput Press, 2004.

Paterson, Lindsay. *Scottish Education in the Twentieth Century*. Edinburgh: Edinburgh University Press, 2003.

Pedersen, Joyce Senders. *The Reform of Girls' Secondary and Higher Education in Victorian England*. New York and London: Garland, 1987.

Petschauer, Peter. *The Education of Women in Eighteenth-Century Germany*. Lewiston, NY: Edwin Mellen, 1989.

Purvis, June, ed. *Women's History. Britain, 1850–1945*. London: University College London Press, 1995.

Raftery, Deirdre, Jane McDermid, and Gareth Elwyn Jones. "Social Change and Education in Ireland, Scotland and Wales: Historiography on Nineteenth-Century Schooling." *History of Education* 36, 4 (2007): 447–463.

Raftery Deirdre and Susan M. Parkes. *Female Education in Ireland, 1700–1900: Minerva or Madonna*. Dublin and Portland: Irish Academic Press, 2007.

Rendall, Jane. *The Origins of Modern Feminism. Women in Britain, France and the United States, 1780–1860*. Basingstoke: Macmillan, 1985.

Robinson, Wendy. *Pupil Teachers and their Professional Training in Pupil Teacher Centres in England and Wales 1870–1914*. Lampeter: Edwin Mellen Press, 2003.

Rogers, Rebecca. "Boarding Schools, Women Teachers and Domesticity: Reforming Girls' Secondary Education in the First Half of the Nineteenth Century." *French Historical Studies* 19, 1 (Spring 1995): 153–181.

———. *From the Salon to the Schoolroom. Educating Bourgeois Girls in Nineteenth-Century France*. University Park: Pennsylvania State University Press, 2005.

———. "Learning to be Good Girls and Women: Education, Training, and Schools." In *Routledge History of Women in Europe, 1700 to the Present*, ed. Deborah Simonton, 93–133. London: Routledge, 2006.

Ruane, Christine. *Gender, Class, and the Professionalization of Russian City Teachers, 1860–1914*. Pittsburgh: University of Pittsburgh Press, 1994.

Schmuck, Patricia, ed. *Women Educators: Employees of Schools in Western Countries*. Albany: State University of New York Press, 1987.

Simon, Frank. "Education." In *Historical Research in the Low Countries*, ed. N.C.F. Van Sas and E. Witte, 58–67. The Hague: Nederlands Historisch Genootschap, 1992.

Sotiropoulis, Carol Strauss. *Early Feminists and the Education Debates: England, France, Germany, 1760–1810*. Madison and Teaneck, NJ, Fairleigh: Dickinson University Press, 2007.

Southard, Barbara. *The Women's Movement and Colonial Politics in Bengal: The Quest for Political Rights, Education and Social Reform Legislation, 1921–1936*. New Delhi: Manohar, 1995.

Spencer, Stephanie. *Gender, Work and Education in Britain in the 1950s*. Basingstoke: Palgrave, 2005.

Stock, Phyllis. *Better than Rubies: A History of Women's Education*. New York: Putnam, 1978.

Tamboukou, Maria. *Women, Education and the Self: A Foucauldian Perspective*. Basingstoke: Palgrave, 2003.

Theobald, Marjorie. *Knowing Women: Origins of Women's Education in Nineteenth-Century Australia*. Cambridge: Cambridge University Press, 1996.

Trethewey, Lynne and Kay Whitehead. "Beyond Centre and Periphery: Transnationalism in Two Teacher/Suffragettes' Work." *History of Education* 32, 5 (2003): 547–559.

Trimingham Jack, Christine. *Growing Good Catholic Girls: Education and Convent Life in Australia*. Melbourne: Melbourne University Press, 2003.

Vlaeminke, Meriel. *The English Higher Grade Schools: A Lost Opportunity*. London: Woburn Press, 2000.

Vroede, Maurits de. "The Catholic Boarding School for Girls in Belgium before the First World War." In *Bildungsgeschichte. Festschrift zum 60.Geburtstag von Franz Pöggeler*, ed. H. Kantz, 313–337. Frankfurt am Main: Peter Lang, 1986.

Watts, Ruth. "Gendering the Story: Change in the History of Education." *History of Education* 34, 3 (2005): 225–241.

Whitehead, Kay and Judith Peppard. "Transnational Innovations, Local Conditions and Disruptive Teachers and Students in Interwar Education." *Paedagogica Historica* 42, 1/2 (2006): 177–189.

Index

CPSIA information can be obtained at www.ICGtesting.com
231276LV00004B/2/P